CALIFORNIA: A FIRE SURVEY

To the Last Smoke

SERIES BY STEPHEN J. PYNE

Volume 1, *Florida: A Fire Survey*
Volume 2, *California: A Fire Survey*

STEPHEN J. PYNE

CALIFORNIA

A Fire Survey

THE UNIVERSITY OF
ARIZONA PRESS

TUCSON

The University of Arizona Press
www.uapress.arizona.edu

Printed in the United States of America
21 20 19 18 17 16   6 5 4 3 2 1

ISBN-13: 978-0-8165-3261-2 (paper)

Cover designed by Leigh McDonald
Cover photo by Kari Greer / NIFC

Library of Congress Cataloging-in-Publication Data
Names: Pyne, Stephen J., 1949– author. | Pyne, Stephen J., 1949– To the last smoke ; v. 2.
Title: California : a fire survey / Stephen J. Pyne.
Description: Tucson : The University of Arizona Press, 2016. | Series: To the last smoke / series by Stephen J. Pyne ; volume 2 | Includes bibliographical references and index.
Identifiers: LCCN 2015035282 | ISBN 9780816532612 (pbk. : alk. paper)
Subjects: LCSH: Wildfires—California—History. | Wildfires—California—Prevention and control—History. | Forest fires—California—History. | Forest fires—California—Prevention and control—History.
Classification: LCC SD421.32.C2 C685 2016 | DDC 363.3709794—dc23 LC record available at http://lccn.loc.gov/2015035282

To Sonja,
old flame, eternal flame

CONTENTS

SERIES PREFACE

To the Last Smoke

W HEN I DETERMINED to write the fire history of America in recent times, I conceived the project in two voices. One was the narrative voice of a play-by-play announcer. *Between Two Fires: A Fire History of Contemporary America* would relate what happened, when, where, and to and by whom. Because of its scope it pivoted around ideas and institutions, and its major characters were fires and fire seasons. It viewed the American fire scene from the perspective of a surveillance satellite.

The other voice was that of a color commentator. I called it *To the Last Smoke*, and it would poke around in the pixels and polygons of particular practices, places, and persons. My original belief was that it would assume the form of an anthology of essays and would match the narrative play-by-play in bulk. But that didn't happen. Instead the essays proliferated and started to self-organize by regions.

I began with the major hearths of American fire, where a fire culture gave a distinctive hue to fire practices. That pointed to Florida, California, and the Northern Rockies, and to that oft-overlooked hearth around the Flint Hills of the Great Plains. I added the Southwest because that was the region I knew best. But there were stray essays that needed to be corralled into a volume, and there were all those relevant regions that needed at least token treatment. Some like the Lake States and Northeast no longer commanded the national scene as they once had, but their stories were interesting and needed recording, or like the Pacific Northwest or central oak woodlands spoke to the evolution of fire's American century in a new way. I would include as many as possible into a grand suite of short books.

My original title now referred to that suite, not to a single volume, but I kept it because it seemed appropriate and because it resonated with my own relationship to fire. I began my career as a smokechaser on the North Rim of the Grand Canyon in 1967. That was the last year the National Park Service hewed to the 10 a.m. policy and we rookies were enjoined to stay with every fire until "the last smoke" was out. By the time the series appears, 50 years will have passed since that inaugural summer. I no longer fight fire; I long ago traded in my pulaski for a pencil. But I have continued to engage it with mind and heart, and this unique survey of regional pyrogeography is my way of staying with it to the end.

Funding for the project came from the U.S. Forest Service, the Department of the Interior, and the Joint Fire Science Program. I'm grateful to them all for their support. And of course the University of Arizona Press deserves praise as well as thanks for seeing the resulting texts into print.

PREFACE TO VOLUME 2

N THE SPRING AND SUMMER of 2011 I conducted two road tours of California, one south and one north. It was a rapid primer on California fire. I couldn't complete everything and returned briefly for a couple of events and interviews. In particular, I failed to traverse the far north, places like the Klamath Mountains and the Modoc. The omission was, initially, a matter of time, and later, of space. I had to hold the text under 60,000 words and could not find sufficient parts to delete to make room for what the north would require. Instead, I decided to add the Klamath to a later survey of the Pacific Northwest. But this traded ecology for institutions, a bioregion for a fire province. In plugging one gap it created another because what makes California's story significant is the way it has joined many biomes under a single administration. Places like the far-north Klamath have to bow to the imperatives of the South Coast. While I can *say* that in this study, I can't *show* it. I'll have to do that task by comparing institutional responses to shared biomes in my survey of the Pacific Northwest.

These regional reconnaissances were conceived as an exercise in fire journalism or, as I like to think of them, as history in real time—they are intended, after all, as color commentary. To them I have sought to bring context, particularly a sense of the fire scene as a historical construct. Since these are not academic pieces, I have not tried to impose the same standard of documentation I would for a scholarly text. Rather, I cite sources where I have quoted passages or stated a perhaps counterintuitive fact, identify and thank those people who hosted or otherwise assisted my efforts, and point out perhaps an especially useful work. A note on sources handles the general references. I know only too keenly the number of places left unvisited and the words unsaid. But we write to genres, as the saying goes, and the virtue of the short, essay-driven collection is also its vice.

In the years between writing my text and seeing it into publication California has recovered economically and stabilized politically, yet a punishing drought has kept the landscape abuzz with fires and primed a fire organization for a potential tsunami of flame. California still dominates the national media and America's unsettled fire scene.

CALIFORNIA: A FIRE SURVEY

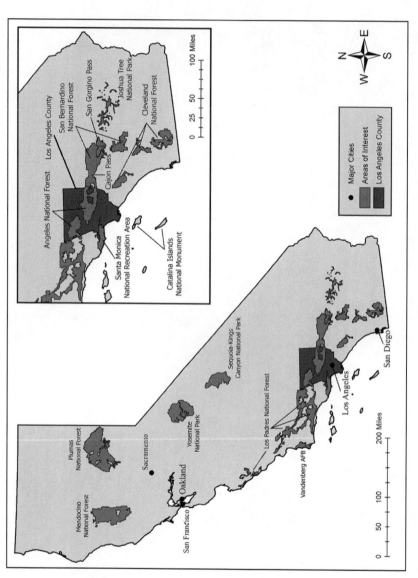

Map of California. Map by Chris Miller.

PROLOGUE

Conflagrating California

The Coming American Fire
Bring me fires to match my mountains;
Bring me fires to match my plains,
Fires with empires in their purpose,
And new eras in their flames.

—ADAPTED FROM SAM WALTER FOSS

CALIFORNIA BURNS, and frequently conflagrates. The coastal sage and shrublands burn. The mountain-encrusting chaparral burns. The montane woodlands burn. The conifer-clad Sierra Nevada burns. The patchy forests of isolated Sierra basins; the oak savannas, on hillsides turning golden in summer; the seasonal wetlands and tules; the rain-shadowed deserts, after watering by El Niño cloudbursts; the thick forests of the rumpled Coast Range; the steppe grasslands of the Modoc lava fields; sequoia, exotic brome, chamise, sugar pine—all burn according to local rhythms of wetting and drying. The roll call of combustible plants and places goes on and on. An estimated 54 percent of California ecosystems are fire dependent, and most of the rest are fire adapted. Only the most parched of Mojave deserts, stony summits, perennial wetlands, and fog-sodden patches of the coast are spared. Not only do fires burn everywhere, but they can persist for weeks and can, from time to time, erupt into massive busts or savage outbursts. Fires can burn something every year. Fire season, so the saying goes, lasts 13 months. Like earthquakes, California experiences a constant background of tremors occasionally broken by a Big One.

This is not news. Anyone even casually familiar with California knows it burns, whether those fires be conflagrating chamise or gas-combusting autos. In fact, most of the United States burns, or can burn, or has burned historically, and virtually every California fire regime resembles those nearby. The northern Coast Ranges and Cascades burn like those in Oregon. The Central Valley is a larger, drier version of the Willamette. The Sierra Nevada looms like a gargantuan Sky Island, the lithic anchor for the Basin Range. The Mojave Desert laps into the Great Basin and Sonoran Deserts. California's shrublands are Arizona's on steroids. Each biota, in brief, has an echo elsewhere. Northern California has lightning fire busts; so do the Cascades. Southern California has extensive burns; so does central Nevada. California slams disparate regimes together; so do most western states. Eastern and western Washington, or northern and southern Idaho, have as little in common as the California's postmodern pastiche.

Rather, what makes California's fire scene distinctive is how its dramatically distinctive biomes have been yoked to a common system and how its fires burn with a character and on a scale commensurate with the state's size and political power. California has not only a ferocity of flame but a cultural intensity that few places can match. In the early years Northern California commanded the scene; after World War II, Southern California; but the state as a whole has a concentrated firepower without parallel elsewhere. Northern California beat back the challenge from light-burning by devising systematic fire protection: Southern California dampened the movement to restore fire by pushing fire management into an urban fire-service model. In the pyrogeography of America, California is the great disturbance in the Force. In ways unmatched by any other region it has projected its presence—its fires and fire practices—throughout the country.

The reasons involve more than bigness alone. In some years Alaska has immense fires, some the size of northeastern states, yet they do not upend national policies. Texas holds more land, and in recent years has been overrun by fires larger in area than those that sweep California, yet those flames have not bonded with a national agenda. New York has a similarly split geography between concentrated metropolis and rural countryside, with the "forever wild" Adirondacks as a wilderness backdrop, yet it remains invisible on the national fire scene. By contrast, California's fires are instantly and hugely broadcast, they infect national institutions, they have repeatedly defined the discourse of fire's history. No other state has so shaped the American way of fire.

Along with Texas and Alaska, it has long behaved as a state-nation. The issue is more than size: Nevada is large, but until recently it grew as an appendage to California; its fire busts pass through the national consciousness with no more effect than wind gusts over a salt playa. The state-nations, by contrast, have unusual political histories, land owner-ship patterns, and creation stories that make them exceptional. Texas was a latent nation for nearly 10 years, although effectively a protectorate of the United States; and when it entered the union, it did not surrender its unpatented lands. A large place with a small population, it never devel-oped adequate institutions, relied on "big men" (large landowners or cattle or oil barons) to run society, and evolved a lively folk culture. Alaska was a territory for 90 years, geographically and politically isolated from the rest of the country, and when it was admitted, it was able to negotiate a division of lands and retain many federal subsidies not available elsewhere. Like Texas a high culture never took root and it had a distinctive origins myth. Unsurprisingly, both voice secessionist tendencies during times of stress. Texas tends to view the United States as France does the European Union, as a means to amplify and project Texas values. Alaska sees the United States as a source of economic subsidy, often behaving more like a country bonded under a commonwealth than like one of the 50 states.

California shares their size, political isolation, and sense of separate selfhood. It began—almost—as a separate country during the Bear Flag republic; by agreeing to accept existing Mexican land grants with the Treaty of Guadalupe Hidalgo, it carved out the choicest lands and created a pattern of large landowners; with the gold rush it has an origins myth founded on a peculiar dynamic of untrammeled individualism and in its early moments of unfettered exploitation. Unlike Alaska it has a fully functioning economy, although one sharply subsidized. Unlike Texas it has never confused size with significance. It has been independent but never secessionist. Unlike Texas, too, it has a bond to high culture; from its onset it has attracted an urban and intellectual society, rich in litera-ture, painting, and later film, which it has reshaped and broadcast back across the country. And unlike Texas it has not exploited the national scene to magnify California sensibilities, but has seen itself as coalescing and shaping a national story, and even an international one. California's is, paradoxically, a cosmopolitan parochialism. All this is reflected in the character of its fires.

California is, in Wallace Stegner's oft-repeated remark, like the rest of America only more so. Today, nearly one in nine Americans lives in California; the California economy is the eighth largest in the world; and depending on perspective California either anchors or weighs down national trends. The California fire scene boasts commensurate dimensions. Southern California holds one of the three dominant wildland fire cultures in the United States; it commands 50 percent of the national fire budget, and an inordinate fraction of the nation's fuels management funds; it claims most fireline fatalities, 35 percent; it has one of two national fire research labs and one of two fire equipment development facilities. Of the 11 coordination centers that make up the national fire dispatching system, two lie wholly within California. That profound distortion is also historical: California was where the debate between fire fighting and fire lighting as alternative national policies was fought out, where systematic fire planning was first devised, where presuppression schemes achieved their most grandiloquent expression, where air attack and the interregional hotshot crew emerged, and where the modern organizational structure for fire—the incident command system—was hammered out before going national, and then global. The California model proved portable in ways the Florida model or the Great Plains model never did. The country is better for that—and worse, as California's recent political dysfunctionality also translates into fire. California is indeed like the rest of the country only more so.

═══════

That there are two Californias is a commonplace. But there are actually many Two Californias; between north and south, Sierra and seacoast, rich and poor, forest and shrubland, megalopolis and wilderness. What makes the state distinctive is the exaggeration that exists between them, the intensity of their contrast. Between Mount Whitney at 14,500 feet and Death Valley (-280 feet), the nation's highest and lowest elevations, is a mere 70 miles as the crow flies. Between Los Angeles and the San Gabriel Wilderness is little more than a line etched on maps. What elsewhere might stretch over hundreds of miles, California has compressed into a relatively narrow chain of mountains, valleys, deserts, and seashore, and in the course of that squeezing it has pushed peaks higher and basins lower, and in shrinking the distance between them has caused the contrast to sharpen and the strength of their interplay to heighten. What is true for its geography is also true of its history. It has evolved in decades what took other countries centuries. In 1848 the Euro-American population was perhaps 15,000; a century later it was 10.5 million and headed to a polyglot 37 million by 2010. The gold rush overlapped with the wilderness paeans of John Muir. What is true for geography and history is also true of society. It grew big suddenly without the nuanced interplay of a networked society; the leap from aboriginal hunter to captain of industry came almost overnight. In place of evolved social relations it substituted technology. And this exaggerated compression is no less true of its fires.

Its natural fire geography aligns along two axes. The axis for Northern California is the Sierra Nevada. It rises highest in the southeast and slants lower to the north and west. Call it the Big Tilt. The axis for Southern California is the Transverse Range. It's highest in the east, where it bends sharply westward, and trails toward the Pacific before making a complementary bend to the north. Call it the Big Kink. Together they make a coordinate geometry within which lies most of what is nationally significant about California fire. Of course there are places in the Mojave that burn, and the Klamath Mountains and California Cascades are subject to punishing lightning fire busts. But neither has shaped a larger discourse. The fires associated with the Sierra massif and with the quirky Transverse have.

The above is an abstract partition of California fire. When formal fire protection began, the California fire community had a different way of parsing fire problems. Partisans divided its two realms of fire along a crenulated edge that separated brush from timber. Wherever that border appeared, foresters concentrated their firefight; they struggled to hold the brush from further encroachment into the timber. The conception acquired a practical expression when, during the 1930s, the Forest Service used the collective muscle of the Civilian Conservation Corps (CCC) to construct the Ponderosa Way, a fuelbreak and road system that stretched the entire length of the montane Sierra Nevada from Bakersfield to Redding. Since the south was mostly brush, and the north mostly timber, in a loose way the biotic distinction reflected a crudely geographic one. Put differently, fire officers wanted to keep the south from overrunning the north. Today, an equivalent fractal border is the intermixed zone between urban and wild, which lies most dramatically in the south but is found throughout the state. After all, the worst urban fire in decades struck Oakland, and the Tahoe Basin is as hopelessly jumbled as the South Coast.

Perhaps the most useful frontier, however, is the one traced by the San Andreas fault zone. South of it lie the fire regimes of Southern California, north of it, those of Northern California. Each zone has its characteristic big fires, its fire culture, and its claims on the national narrative. But it is where the San Andreas suddenly veers west from the north-trending Peninsular Range to raise up the Transverse Range and define the Los Angeles Basin that the border hardens. The San Bernardinos, the San Gabriels, the Santa Monicas, and the San Rafaels that collectively trace the Transverse Range have no equivalent elsewhere in the nation. To the far south, in the Sea of Cortez, the tectonic blocks of Earth's epidermis spread laterally, pulling crust apart to make a new ocean, and to the far north, it turns into a subduction zone, pushing one plate below the other. But along the San Andreas, a literal transform fault, the two giant plates of Earth grind along, muscling up mountains by wrenching and lifting, and where the fault abruptly veers west for some 250 miles, the strains increase and the mountains rise more sharply into the massifs of the Transverse. Where the fault kinks again, above Santa Barbara, it forms a kind of triple junction, flaring northwestward to the Coast Ranges and northeastward into the Sierra Nevada.

So it is, too, with California's pyrogeography. The fire province that is Southern California is both isolated and self-contained. To the south lies Mexico. To the east is the Peninsular Range. To the north runs the Transverse Range. To the west stretches the Pacific Ocean. It's a geographic box, in its way as distinctive as Florida. The big kink defines a tectonic shift as profound for fire as for geology. It slams a mediterranean climate against foehn winds: a shrub-choked, sheer-sloped landscape meets some of the most savage winds on the planet. Fires burn one way north of the line and another south. Measured by the number of fires, the big busts lie north; by fire size and intensity, they lie south. This distinction even holds for urban fires. Where the San Andreas Fault plunges into the Pacific, it widens into a suite of faults along the San Francisco Bay Area. Where it kinks to make the Transverse Range, the largest metropolis in the United States washes onto its flanks. California's Big Ones break out along the San Andreas; that is also where most of the population resides—where some of the country's most aggressive urban sprawl abrades against some of its most savage fires. The Big Kink makes California fire unique and uniquely powerful.

These are the geographic facts. The political fact is that both realms reside within a single state: the politics of reconciling north and south means that both cultures will alloy. What is curious is that, repeatedly, while national ideas have come from the north, it is in the south that they have been institutionalized and have gelled into a culture. More people live in urban clusters, after all, than in wilderness. By 2000 the urbanized South Coast held some 8 percent of the state's land and 56 percent of its population. Accordingly, the setting south of the San Andreas has exerted institutionally a political force that created a Big Kink in California's fire scene, which has then distorted the national scene. The process threatens to transform a California singularity into an American norm.

———————

California's role as a national innovator in conservation arrived early. It created a geological survey in 1860. It founded the University of California in 1868. It established a Board of Forestry in 1885. It looked to the federal government to oversee hydraulic mining with the Debris Commission. It boasted the second cluster of forest reserves established in the country. Even before the Forest Service took over those reserves, timber owners in Northern California were experimenting with fire prevention, and associations in Southern California were carving fuelbreaks across the mountains and sending fire guards to patrol them. The U.S. Bureau of Forestry pointed to both experiments as national models. In 1905 the Bureau assumed control over the forest reserves, which it redubbed national forests, and then renamed itself the U.S. Forest Service. The California experiments became national exemplars.

That pattern continued throughout the 20th century. When the Far West debated conservation—was dividing between untrammeled logging, mining, and ranching and protected public lands, and was debating what fire policies were appropriate for each—both sides staked their claim in California. One favored a perpetuation of frontier practices, distilled into the phrase "light-burning," which they promoted as "the Indian way of forest management." Regularly burn the montane woods and lowlands, they insisted, and you would not have ruinous fires. The other side favored active fire protection by applying the science of the day with the force of government authority to prevent fires, fight fires, exclude fires. The light-burning controversy was a national issue, but it was argued in California, and in 1923 a commissioned board investigated and formally anathematized the practice. The California conclusion, encoded by academic foresters and government bureaus, washed over the rest of the country. Revealingly, California not only denounced the opposition. It defined, in an aggressive way, the active alternative.

The 1910 fire season burned away naïve notions that progressive ideas and simple good will would solve the national fire problem. In particular, the Big Blowup that swept over the Northern Rockies in August traumatized the young agency and, combined with a challenge to policy posed by light-burning, hardened its conviction that fire protection was the foundation of its administration. But the means to make it happen came when ranger Coert duBois, after struggling with fires around Tahoe, sketched a plan for organized fire protection in 1911. Three years later, as Forest Service chief for all of California, he published the classic *Systematic Fire Protection in the California Forests*, in which he adapted the operations engineering of his day to the problem of fire control. No other scheme had anything like its rigor or comprehensiveness; over time, it became the basis for wildland fire plans nationally.

The duBois project demanded not only data but the means to apply its conclusions. In 1919 duBois serendipitously met Captain H. H. Arnold of the Army Air Corps in a San Francisco saloon, out of which arose a program of joint aerial reconnaissance that marks the onset of air attack. Meanwhile, the purpose behind those practices received support from systematic fire research under the direction of S. B. Show, bolstered by collaboration with E. I. Kotok; and although Harry Gisborne in the Northern Rockies was a far more colorful character, the remorseless Show and Kotok program fed directly into duBois's schema and defined the main thrust of organized research. In 1921 the Forest Service convened its first conference on fire protection at Mather Field, outside Sacramento. The 1924 California fire season, about which even pulp novelist Zane Grey wrote hectoring denunciations, inspired a national board of review—another California innovation for instilling rigor. What was a periodic crisis in the Rockies and a chronic irritant in the South had yielded, in California, to an organized program.

The following decades only bolstered that trend. Fire protection became a massive public-works project during the New Deal. California claimed a lion's share of CCC camps and other emergency resources, and it undertook the most ambitious projects in presuppression. (If the 1932 Matilija fire was a forestry equivalent to the Dust Bowl, the Ponderosa Way fuelbreak across the western Sierra Nevada was a fire equivalent to the Great Plains Shelterbelt.) It fielded the first mobile weather units, complete with a fire meteorologist. It established the first genuine investigations into fire behavior, combining field trials, lab experiments, and even wind tunnels. Although it stripped away many firefighting resources, World War II strengthened the political claims of fire protection: it militarized fire defense as national defense. A Southern California alarmed over possible Japanese attacks convinced the Wartime Ad Council in 1942 to create the first national program for forest fire prevention, which by 1944 morphed into Smokey Bear. The war response suggested how military hardware could move into the backcountry. A Cold War on fire made it happen.

By then, too, the postwar migration to California was under way. The economy boomed, partly because of its natural wealth and agriculture, partly from construction, and partly from continued massive subsidies from the federal government, primarily military (the Department of Defense oversees 5.4 million acres, but the navy has two-thirds of its investment in California). It had the effect of a new gold rush, almost exactly a century after the original. So powerful was the shock that it rewired the nation's geography. When World War II ended, no professional sports team was west of the St. Louis Cardinals. Before the 1950s ended Los Angeles had the Dodgers and San Francisco the Giants. The country was fast becoming bicoastal. California dominated the West Coast, and through it, everything west of the Rockies.

Power followed, not least power to influence the national fire establishment. Again, California became the center of national trends, and as had been the case 50 years previously, the debate over both the means and the ends of fire protection revived the California split. On one side—the south—urban sprawl remade the rural landscape and effectively shifted fire protection to the model of a city-centric fire service. This was a vision of fire abolished or at least contained. On the other side—the north, specifically the Sierra parks—came a very different vision of fire, one of fire restored rather than removed. It argued that future generations would need to reinstate the fires that previous generations had unwisely wiped away. These were incommensurate perspectives and mutually impermeable. They reflected, in fact, the splitting of the postwar landscape, and it is no accident that this controversy, like the one about light-burning that preceded it, found its center of gravity in California. The American fire revolution that followed is largely a chronicle of how these competing landscapes, their fires, and the institutions that oversaw them have worked along that fiery fringe.

It happened first in California. In 1961 what had been confined to outliers like Malibu or Wheeler Canyon moved into the Los Angeles city limits as brush fires raced along the flanks of the Santa Monicas and blasted through such exclusive tracts as Bel Air and Brentwood. They gave the media its new set piece image of fire Americana—some ill-clothed doofus on a combustible roof with a garden hose. The next year Interior Secretary Stewart Udall commissioned a committee to report on the festering sore of elk at Yellowstone. Chaired by Starker Leopold, a wildlife professor at the University of California, Berkeley (and son of Aldo), the committee expanded its charge, and trekked to California. There it argued that America's parks should stop being recreational theme parks and aspire to be "vignettes of primitive America." They should allow fires to return, as though they were wolves and grizzlies, and permit nature to work out its own ecological destiny as little fettered by human artifice as possible; or where the insults had gone on for too long or scoured too deeply, those damaged processes should be rehabilitated and restored. Out of the 1963 Leopold Report came a complementary iconic image, that of friendly flames lapping along the base of the Big Trees. Its Sierra parks hosted California fire's counterculture—the communes of back-to-nature folk committed to reinstating an imagined former world. Between them the shake-shingle roof and the Sequoia grove defined the polarities of the emerging fire scene nationally.

The Leopold Report, the Wilderness Act, and the Tall Timbers fire ecology conferences (which began in 1962) converged to transform fire control into fire management, and to bond fire in principle to land management. All that was fine for a public estate dedicated to nature preservation; wilderness, after all, was a fight over the future of existing public lands. And it was suitable for private landowners, almost all in the Southeast, who wanted traditional working landscapes and sought traditional means to sustain them. But already fire as part of working California landscapes was dying out for ranchers and loggers, and while vast patches of the public domain were gazetted into wilderness, the main arena for such fires moved to Alaska, the Northern Rockies, or such remote ranges as the Southwest's Mogollon. The reforms said nothing about the most explosive category of American land use, urban sprawl, whether it be edge city, suburb, or exurb.

For those lands fire management meant fire suppression. California was to sprawl as Alaska was to wilderness: it expressed that landscape in its purest, most expansive, and most exaggerated form, all shouldering against an extraordinarily fire-prone fringe of mountains. Everywhere, sprawl interbred with natural hazards, but these were rare like hurricanes, or small in scale like tornadoes, or large nuisances like floodplain overwash. But wildfires in Southern California that shot flames and rained embers on cities were violent, photogenic, and distressingly frequent. They were unavoidable. They seemed to require a counterforce equally powerful and implacable. Southern California responded by fielding an industrial-strength firefighting force.

It mutated into a more virulent form by mating with the kind of machines that had won the war. In 1945 an equipment development center was established at Arcadia to midwife that ambition. Its most visible expression was—again—aerial attack. As the Korean War wound down, fire authorities promoted a dedicated program, Operation Firestop, to transform technology into operational programs; the experiments were staged at Camp Pendleton. By 1956 aerial tankers were dropping retardant along California firelines. Although smokejumping remained an artisanal practice in the Northern Rockies, aerial firefighting—mechanical force applied on a systematic basis—was as much Californian as the state's postwar aerospace industry. In 1961 its specialty fire crews (hotshots) went national, now available for duty outside the region. In 1963 the Forest Service opened the Western Fire Lab in Riverside. Its themes were fire behavior and fire planning—essentially the duBois and Show program—bolstered by air attack. Its showcase project, however, was Firescope, which after the 1970 fire siege sought ways to coordinate the maddening complexity of jurisdictions and fire agencies in Southern California toward common practices on the line.

As the fire revolution matured, various regions identified their specialties. The Southeast, keystoned by Florida, became the national oracle for prescribed fire. The Northern Rockies became the authority in the management of fire over a vast backcountry. And California, notably Southern California, became the national center for fire suppression. Every aspect of fire management bent to that imperative. Equipment development labored to make better tools to fight fire. Research explored fire's behavior as a hostile force, fire's management as a problem in fireline operations, fire's effects as threats to urban life. The concept of the wildland-urban interface came from Southern California. So did helitack, air tankers, hotshot crews, wildland fire engines, the incident command system. Southern California is to wildland firefighting what Hollywood is to movies. While elsewhere fire management became a more complicated component to land management, here the complex distilled into a single imperative: to protect assets from fire. Land management distilled into fire management and fire management into fire suppression. Elsewhere fire management was one of many tasks for land managers. Here land management was a series of tasks to support fire protection.

As the firefight intensified, the casualties mounted. There had been serious breakdowns before the war. The Griffith Park fire of 1933 killed 25, and the 1943 Hauser Creek fire burned over a company of marines, killing 11. But these were not crews trained for fire duty. The postwar era saw burnovers spread as fast as suburbs. The 1953 Rattlesnake fire killed 15 firefighters. The 1956 Inaja fire killed another 11 and sparked a reform in fire training. The 1966 Loop fire burned over the El Cariso Hotshots, killing 12 and reinvigorating training. The 1968 Canyon fire killed eight. By then more firefighters were dying from accidents involving internal combustion than from flame. That burden shifted to civilians living, literally, in the line of fire. Some 25 died in the 1991 Tunnel fire, and 15 in the 2003

Cedar fire. The scene was going rogue. Each outbreak brought a swarm of engines, crews, and aircraft from around the state, and often, the nation, only to be overwhelmed by the next outbreak. By 1993 Southern California fire had become part of the news media's almanac of natural disasters. What Joan Didion imagined as Los Angeles's "deepest fear"—the "city burning"—was no longer a conspiratorial threat or a literary trope but an annual ritual. By the early 1990s wildfire had joined other social and natural disasters to make Southern California, in Mike Davis's wording, the epitome of "the imagination of disaster."

The California fire scene had long shaped the national agenda. That it has traditionally treated the neighboring West as a kind of colony, and the fact that these states hold the bulk of the public domain, has granted it additional leverage over the national fire narrative. In 1964, when California still dazzled the rest of the country with its promise, and Lassie had joined the U.S. Forest Service, California fire seemed a portal to a golden future. Nearly all Forest Service regions had Californians as fire officers. As the 20th century ended, however, shaping the narrative had morphed into distorting it, and the Beach Boys had given way to *Blade Runner*. It threatened to become a maelstrom sucking everything into its vortex. Fire agencies elsewhere bristled as they saw equipment, crews, money, research, and policy sucked into the California maw, with no measurable result. In the national mythology California had undergone a reverse alchemy from gold to lead; it was what the rest of the nation's fire establishment did not want to happen. The Ugly Californian became a stock figure of national fire lore. Still, on the 50th anniversary of the fire revolution, the fire directors for both the U.S. Forest Service and the National Park Service were Californians.

═════════════

For all its distinctiveness California fire is of a piece with California's air, water, and soils. All have undergone radical transformations, occasionally in massive ways. Hydraulic mining had moved some six times as much earth as the Panama Canal, all of it spread over the Sacramento Valley or dumped into San Francisco Bay. California has moved its waters on a comparable scale, damming and diverting it from the north to the south; Southern California notoriously drained the Owens Valley, and drafted water from the Colorado River. By the 1940s Californians had so befouled their air that it became an undeniable public health hazard; the Los Angeles Basin had the most polluted air in the country, and the Central Valley was not far behind. What Californians did to fire—disaggregating, relocating, erecting massive infrastructure to cope with the consequences—was little different. As with its waters California sucked fire apparatus from north to south.

Everything was done to support economic development, of which the postwar housing boom was only the latest iteration. But it was also done without the buffers of a rooted society. A wave train of immigration, each wiping away its predecessor, had bequeathed a society not bonded by long association and patient labor to its lands. It confronted nature—an exaggerated, panoramic nature—directly. Its fire practices did not reflect centuries of trial and error. Rather, they sought to protect the latest economy from fire's threats, or they sought to leave nature alone with its flames. There was no organic unity to California's split fire scene because there was no social order to hold them under a single valence.

This lack of a traditional fire culture pointed to structural rather than behavioral solutions. It argued for fuelbreaks rather than prescribed burning, for fire engines rather than prevention, for building codes to shield houses rather than zoning to prevent construction in the first place. The Southern California scene, in particular, was haunted by the specter of arson—arson not so much antisocial in its character as asocial. It spoke to the lack of a coherent society, one whose confusions and contradictions were being manifest on the mountains. Increasingly, there seemed no middle ground: the successor to light-burning was let-burning. When flames appeared the options were fight or flight; there was no ongoing engagement rooted in common values of land and culture.

For the national narrative the important fact is that what happened in California did not stay in California. Ideas immigrated to California, underwent a transformation, often a radical simplification, and then emigrated outward. For many regions of the United States one can tell the local fire story with minimal reference to the country overall. That isn't possible in California. Its story is one in which national themes are acted out, and then rebroadcast over the country. California is the Big Kink in America's narrative of fire.

CALIFORNIA'S INVENTED
FIRE CULTURE

THE GOLD RUSH remains both spark and paradigm for California's American era. A wild surge of polyglot peoples that overwhelmed previous land use, an adolescent habit of cheerful carelessness, a disregard for government institutions and social bonds in preference for self-absorption until crises become overwhelming—these were the defining traits of California society as judged by Josiah Royce, son of forty-niners who became a professor of philosophy at Harvard, when he wrote his classic *California: A Study of American Character* (1886). The Californian showed many of the best features of the American character, and many of its worst, but always exaggerated, unbuffered by history, tradition, and complex social bonds. "Ever since," Royce believed, California has "preserved a curious likeness to the fortunes of the early days, and that, in numerous and recent instances, have led to a more or less noteworthy and complete repetition of certain early trials, blunders, sins, penalties, virtues, and triumphs."[1]

Unsurprisingly, Royce opened his social history of San Francisco—then the major urban and economic center—with a chronicle of Great Fires that not only materially wiped out and renewed the city but symbolize the zealous insouciance and disdain for traditional society that characterized those times and by implication have imprinted themselves on successor eras. One conflagration swept through the hastily erected campsites—"city" seems too strong a term—in December 1849. A second struck on May 4, 1850. A third raged through in June, followed by another in September. On May 4, 1851, a fourth, and greatest, arrived in a macabre commemoration. That June the fires migrated up the surrounding hillsides. In less than 18 months the city had suffered through four major wildfires, which seemed excessive even by California standards.

Still, there remained a stubborn preference for wooden housing and a reliance on "the energy and ability of the large and efficient fire department of the city as one's security against all fires." Tellingly, an outcry arose from the commercial interests "not to pass at once any ordinance restricting or forbidding the building of frame houses within fire limits, since such a measure at that moment would drive away too many who are now hesitating whether to risk another trial of their fortune in the city." Keep the rush going. If matters soured too far, then a Vigilance Committee—in this case, of fire officials—would eventually step in.[2]

A century later the dominant urban scene had migrated to Southern California, but the dynamics remained the same. The reason fires of a similar type returned was because the pattern of land settlement was repeating. And the reason land use recapitulated itself was because the same character of settlement kept renewing itself. Wildfires were the cost of keeping the rush going. Curiously, Josiah Royce devoted his intellectual career to a search for, and validation of, community and the loyalties that weave a society together. For all his exuberance, the Californian, he concluded, "has too often come to love mere fullness of life and to lack reverence for the relations of life." Against that wanderlust and boisterous indifference, Royce sought to place the relations of loyalty at the core of human happiness. He never succeeded. In California the fires succeeded one another.

—————

The reinvented life is an old California trope. At the time of the forty-niners it meant, for individuals, a dramatic break with the past, a severing of generational ties; but it also meant a chronic turmoil, even breakdown, of the new society that simmered in the cauldron of these relocated souls. The process of shattering indigenous society commenced in the 1760s with the establishment of a mission system from Mexico. The gold rush completed the task by introducing disease, violence, land theft, and general mayhem. Within 50 years the native peoples of California, a group that likely had constituted as high a percentage of the pre-Columbian population of America as California does for the America of today, were a handful of survivors scattered like seed among the stones. Virtually overnight, the old order vanished.

There is every reason to believe that indigenous California was a land massaged with controlled fire, not unlike the pattern of "firestick farming" typical of Aboriginal Australia. To these regimes nature from time to time added lightning busts or eruptive winds; but for thousands of years people had lived on an exceptionally fire-prone land and evolved appropriate fire practices. The range of reasons were the usual ones: to improve hunting and travel, to protect against wildfire, to encourage useful shrubs and edible forbs. Some fires would escape. Some—hostile fire—would be turned against enemies. In 1885 Joaquin Miller reflected on a California boyhood spent "with Indians."

> They were the only true foresters I ever knew. In the spring after the leaves and grasses had served their time and season in holding back the floods and warming and nourishing the earth, then would the old squaws begin to look above for the little dry spots of headland or sunny valley. And as fast as dry spots appeared they would be burned.
>
> In this way the fire was always under control. In this way the fire was always the servant, never the master. And by the time the floods came again there was another coat of grass and leaves, stronger and better than the one before, because of the careful and temperate fire of the careful and wise old woman. By this means the Indians always kept their forests open, pure and fruitful and conflagrations were rare.
>
> . . . Let the foresters of plain, hard, common sense follow the Indian's simple method of preserving his property by burning the leaves, and my word for it, neither New York, Louisiana, Michigan nor California need fear flood or fire, drought or drowning rains.[3]

The Mother Lode country over which the forty-niners sprawled, and admired when they weren't ripping it apart, was a landscape so routinely burned by its inhabitants that explosive fires were rare. All that vanished within a handful of years.[4]

The demographic surge has been astounding, and relentless. The 1850 census counted the "historical population" at 92,597. By 1860 that number was 379,994, an increase of 310 percent. For the next three decades the increase was nearly 50 percent by each census. Then it paused at a mere 22 percent. Up to 1930 the increase ranged from 44 percent to 66 percent by decade. The Depression halved that number, though it drafted in a surge of Exodusters. The postwar boom raised the number to 50 percent again. In 1970 there were 20 million Californians. In 2010 that population had nearly doubled to 37 million.

What astonishes is not merely the rate of increase but the proportion due to immigration. The experience of the indigenous Californians would be repeated over and over, as waves of new immigrants overwhelmed the old, and shook the landscape free of former relations. A few communities remained deeply rooted in niche settings, like redwoods along deep-shaded streams. Most, however, resembled California's grasslands, annuals maintained by continual reseeding, full of imported weeds, although most of the exotics eventually naturalized, their origins forgotten in the scramble to work out a future. Reform of immigration law in 1965, and a breakdown in the Mexican border, sparked the largest wave of immigration in the nation's history, along with a shift in demographics from Europe to Latin America and Asia. Much of that flow passed through California.

From the time it entered the United States, that is, California has been a constitutionally unsettled society, one that has sought solutions that did not depend on a sense of common ancestry or shared ties to the land. Fire and movement replaced blood and soil. Land was either another commodity like nuggets or it was something untouched and distant, subject to wonder rather than use. Hydraulickers rubbed shoulders with John Muir. In California farmers called themselves growers, homesteaders behaved like realtors, and government devised means to control fire without relying on ancestral ties to the land or fire traditions passed down through generations. The reinvented life had its counterpart in the reinvented landscape. When the landscape bifurcated into the wild and the urban, the intellectual chasm between them widened even as the geographic distances that actually separated them shrank. Landscape fire split along the same veins. It changed from a routine habit to either an urban threat or a natural marvel. It became peculiarly dissociated from human agency, an ineffable visitation that appeared like earthquakes and Santa Anas.

California brought to the Union 100 million acres. Perhaps 9 percent of this total had legally been granted by Spain, and then by Mexico, to missions and private individuals, the estates often of vast dimensions. Mexico secularized the missions in 1833, releasing those lands. But the experience imprinted on California a pattern of large landowners and masses of workers. Ranchers, timber owners, railroads, growers—the formula persisted. By the onset of the 20th century large swathes of the public domain were excluded from private purchase, establishing another regimen of big estates, this time under the jurisdiction of the federal or state government. By the onset of the 21st century some 52 percent of California was public land. Private lands were rapidly transfiguring into housing tracts, on which resided the new working class for big corporations.

It had all happened very quickly. With astonishing rapidity California's minerals, timber, pasture, soils, and waters were monetized, often with scant concern for environmental aftershocks. The process required huge infusions of capital, which argued for state sponsorship or big business like railroads. Mostly, the federal government, directly or indirectly, installed the infrastructure. It happened on a scale commensurate with an emerging economy; California capital in turn shaped much of the western United States. And it occurred at a tempo that, combined with the influx of newcomers, made any familial ties to the land impossible. Inherited land uses blew away with the next wind of economic enthusiasm; owners sold rather than passed down from generation to generation. California was neither a blooded nor a rooted land.

Nor, as Josiah Royce observed, was that land bounded by social duties. The economy boomed through massive public works that did not evolve out of long affiliations and organic trial and error. The footloose population was inclined to look to new technologies rather than inherited ones. It locked up criminals at a rate higher than elsewhere. It sequestered its insane at phenomenal rates. Rather than neighborhood policing, it looked to technological surrogates—patrol cruisers and helicopters, for example, rather than community relations. Its solution to rising crime was to commission more prisons. Its response to chemical pollution and carcinogens was to ban outright. It ideally solved chronic land abuse by leaving land alone entirely. Rather than rely on social norms to regulate behavior, the state erected a rigid infrastructure as a shield or it suppressed what was deemed dangerous. Hard-power investments by government replaced the soft-power ordering of society. That half its landed estate was private gave California a freedom not available to Alaska. That the other half was public lent a counterweight not available to Texas.

It should be no surprise then that California's fire history displays similar traits. California repeatedly rejected common lore, which it lacked, in favor of novel technologies, ideas among them. It erased indigenous fire practices; it repudiated landscape burning; it denounced folk traditions. Instead, it depended on big institutions, on the equivalent of wealthy landowners (such as counties), on extensive public works, on a federal largesse, and on either abandoning or locking up what it regarded as deviant. It protected society from fire as a mechanized constabulary did from crime. A casual perspective would suggest that California did not—does not—have a fire culture in the sense of fire practices bonded to land. There is nothing like the preserved lore characteristic of the Red Hills of Florida or the Flint Hills of Kansas. Contemporary California fire norms have no more in common with pre–gold rush habits than casinos do with basket weaving.

Yet California did create a fire culture to match its mountains. It happened where the mountains were, on the public lands. It was not a bond to the land so much as to the institutions that oversaw the land, notably the U.S. Forest Service (USFS). The USFS didn't invent forest firefighting, of course, but it did remake it into *systematic* fire protection, and it was California that hosted the demonstration plots, fought off folk traditions of light-burning, and forged a bureaucratic task into something like a way of life. These norms were passed down not only to new recruits but to the sons (and later, daughters) of those who lived them. Some fraction of those who grew up in national forests and parks would themselves join the agencies; there were second- and even third-generation offspring, not unlike military families. Their loyalty, their learned culture, was not to a particular land so much as to the agency.

In *Young Men and Fire* Norman Maclean conjured a scene in which the young smokejumpers walk into the Mann Gulch fire, each imagining himself in a stadium with his father, "who fought fires in his time," in the stands watching. In California that was often literally true. The Millars, the Biddisons—sons replaced fathers as fire officers. In the Southeast burners still speak of how they learned their art from their fathers and grandfathers, who took them out to the fields with torch in hand. In the West fire officers could invoke episodes where their fathers taught them to stomp out campfires or swat out small blazes. Chuck Mansfield recalled an episode when he was seven and his father took him to search out a smoke that lookouts kept reporting but no one could find on the ground. They eventually smelled their way into it, and then "Dad cut off a branch from a young fir and showed me how to use the branch to beat out the fire." (When he grew up, Mansfield signed on for a tour with the smoke-jumpers.) For decades, seasonal firefighting was an entry level position for all the federal agencies. It was a rite of passage for virtually everyone. Fire became part of agency culture, which evolved into a species of fire culture.[5]

It could be a way of life. Consider Ron Watson, retired California Department of Forestry fireman. His extended family constituted a fire clan. His brother retired as a dozer operator. Another brother worked in air attack. His wife, sister-in-law, and their twin boys had been fire-fighters. His bother-in-law was an engine captain. His son-in-law was a captain who trained inmate fire crews. He had in-laws who served as fire lookouts. Even by the norms of California, even for a fire brotherhood that can resemble a caste and a wildland mission that hybridizes with the urban and has absorbed some of its traits, this passed beyond the routine. What may be most interesting, however, is not its demonstration that wildland fire mattered and that fire could animate a subculture, but that the fire practice that mattered was suppression.[6]

The process commenced when new landowners searched for fire practices to replace those extinguished with the indigenes. Sensing the chance to begin anew, foresters argued for a fire administration that would be based on first principles as articulated by their science. Fire officers saw frontier fire habits as part of a package of bad practices which reserved forests were established to halt. Fire control was not nuanced: it was a universal goal intended to provide a common standard over a rudely rural and locally wild landscape. In the postwar era that landscape broke apart, like tectonic plates widening to make the Gulf of California. The fissuring split fire management.

One side committed to development. It began to log off the reserves and it crowded the borders with housing tracts. Everything in the south, especially, bent to support the continued expansion of the regional economy, and when that economy turned to urbanization, an apparatus for fire protection developed as prominent as that for water delivery to field and townsite. Over and again, an urban fire service had to grapple with brushy hillsides and banshee mountain winds, while wildland agencies had to move from logging slash to single-family houses; and because, under extreme conditions, no single agency could cope with the demands made upon it, each had to cooperate with the others for the duration of the emergency. The experience forced public agencies to shrink their options into preparing the land for fire, fighting fire, and rehabilitating sites after fires. Whatever their intention, however much they might imagine themselves as foresters or land managers, they became a support service for sprawl. Some of the most explosive fires on Earth fronted some of the most insistent urban growth. A hybrid, high-octane firefighting agency was the result.

The other side of the chasm sought an ever-purer wildland. The Leopold Report for the National Park Service and the Wilderness Act for federal lands generally made the presettlement era before the gold rush California's true golden age. By 2000 some 14.3 million California acres were legal wilderness and much of its national parks was managed as such (by contrast, the state's urban footprint was a third as large). Land reservation on this scale required more than an explanation; it required a narrative. It's not hard to see why and how a California rich in imported intellectuals and a monumental natural setting might invent traditions to create a past it had erased. The idealized fabrication of the Spanish mission era is a good example. (The theme park further distilled the process for maximum effect.) Even in forty-niner days California had a vibrant literary tradition with national reach. In the 20th century it added film and TV.

As its Santa Barbara–based historian Roderick Nash has observed, wilderness is an idea, a state of mind, not a reality or state of nature. Here was another California creation dissociated from lived-on land: the whole premise was that the land had no permanent residents and hence no inherited lore. What fire culture might exist would have to be contrived out of whole cloth, and paradoxically it would have to deny that it was a culture at all. In its yearning for a prelapsarian past, in its reliance on literary construction, in its appeal to the elite and upper-middle class, in its appearance during a period of industrialization, in its attraction to reenactors, Wild California is of a piece with Highland Scotland.[7]

When California's landscape split, so did its notions of how to manage wildland fire. Urban California adapted the city's concept of fire protection, and flung it over the near backcountry. Wildland California sought, by stark contrast, to remove the human presence, most visibly by staying the hand of fire suppression and expanding the realm of free-burning fire to the city's edge. In both enterprises it helped that California, while lacking an organic culture, had from the onset of its American origins a lively literary and artistic one. If California lacked a rooted society, it could create a magnificent simulacrum or surrogate.

Whether or not its imagined world is historically true is irrelevant. It has become true because people have acted on the beliefs it embodies. Legal wilderness and national park backcountry are a reality; fires occur within them and must be managed. Unsurprisingly, California is where much of the impetus and techniques have come from. California didn't invent let-burning any more than it did light-burning or fire suppression. What it brought to the topic was system and state power. An invented fire culture became serious and nationally influential.

⸻

In the National Cohesive Wildland Fire Management Strategy proposed in March 2011, the federal fire agencies identified three primary concerns. They sought resilient communities, restored biotas, and the capacity to fight unwanted wildfires. Or put into fire terms, they needed to fight bad fires, light good ones, and leave more room for natural ignitions to work their ecological alchemy. These ambitions equated with three types of fire. Prescribed fire was deemed indispensable to restore damaged (and fire-famished) ecosystems. Natural fire seemed the most appropriate way to reinstate free-ranging flame into untrammeled wildlands. And fire suppression continued to be the sole option for protecting the built environment from unleashed flames. Expressed geographically these belonged, respectively, on the working landscape, the wilderness, and the wildland-urban interface.

It is worth noting that all had their histories bent, like light passing around a black hole, by California. Light-burning suffered its greatest political defeat in Northern California, effectively banishing it to the outer darkness of Florida cowboys, Native American tribes, Osage County ranchers, and prairie partisans. Restorative burning, particularly through natural fire, evolved out of the Sierra parks. The mutation of smokechasing into I-zone (wildland-urban interface) suppression achieved its most remorseless expression south of the San Andreas Fault. Here, wildland fire management became an urban fire service by other means. Those who yearned for a different direction for American fire management might be forgiven if they muttered about a California curse that tolerated California's oversized influence when it was golden. Now that it had turned to lead, spreading like sudden oak death, they wanted to quarantine it lest the political paralysis that made California seemingly ungovernable contaminated the national scene.

The grand stresses, almost tectonic in scale, that underlie California fire continue. The mountains rise, the cities press against and around them, the fires kindle and blast through the wild and the urban and the sloppy and volatile border they share. Those conditions won't change. There may be no regional solution except to reduce the damages and costs of the relentless friction and the fires they spark. Still, this is the American scene exaggerated. The frightening aspect of the California split is that it is dissolving the middle ground and letting the extremes shout across a narrowing divide no more impermeable than a fuelbreak down a ridge in the San Gabriels. The center cannot hold because, like California's native grasses, it never put down perennial roots and instead relied on continual reseeding. Responses are less informed by fire triangles than by fire triages.

California's landscape, in brief, looks a lot like its society, politics, and legislature, as they become ever more explosive and less manageable. A danger for California is that its lands will in fact emulate its politics. A danger for America is that California will become what its boosters have always insisted it is, a premonition of the future, that California's endless experiments will spill over its borders and become not warnings but prospectors actively sluicing out a future.[8]

STATE OF EMERGENCY

BIG, BURNING, BOISTEROUS—that California should have formidable firepower was always more or less a given. The state is too large and its fires too prominent to ignore. But the shape of state-sponsored fire management was far from foreordained. Local authorities have fire responsibility for roughly a third of California; the federal agencies, for another third; and the rest falls to the state. But it was never obvious how these entities might unite and certainly not inevitable that the State of California would establish an in-house firefighting force and extend urban-style fire services throughout some 31 million acres on a scale that has made it the biggest of its kind in the world, the third largest fire agency in the United States, and a gravitational disturbance to the national commonwealth.

All states have foresters, and all assist with fire protection beyond cities and the federal estate. Many, like New York's, were created to staff state parks or forests. Texas established a forestry bureau, which came to assume rural fire protection responsibilities (or oversight for volunteer brigades), but by 2010 (and keeping with that state's anti-institutional instincts) its permanent staff numbered a scant 375. Alaska's Division of Forestry manages 20 million acres of state forests and furnishes fire protection for 150 million, but has little presence outside those lands. Probably the closest cognate is Florida's Forest Service, which maintains general rural fire protection for most counties and became a national pioneer in institutionalizing prescribed fire. But none compare in scale and heft to CalFire.[1]

Some quirk of California geography and life allowed a puny Board of Forestry to bulk up into a behemoth and have its fire obligations dominate the others. The simplest explanation is the spasmodic tempo of California history, specifically its proneness to cataclysms. No single institution could by itself cope with such catastrophes; each new crisis forced the state to fill in the gaps between other fire agencies and to seek alliances among them; each cataclysm boosted overall capacity a quantum level. So effective did the response become that the system not only reacted massively to cataclysms but paradoxically discovered it could not survive without them. It found new ways to live beyond its means because, in California, the mean has meant little. It is the extreme event that drives history. California became a permanent state of emergency.

In 1885 the state created a Board of Forestry. It was timely if toothless, part of a false dawn across North America to halt untrammeled slashing and promiscuous burning. The Board was able to hire a few "agents" and enroll citizens to help enforce laws about starting and fighting fires, with little effect. It withered away by 1893.[2]

In 1905, the same year the national forests were transferred to the Bureau of Forestry, renamed the U.S. Forest Service, the California legislature passed the Forest Protection Act. The act reconstituted the Board of Forestry, appointed a state forester, and granted him authority to designate volunteer fire wardens who could, in turn, enforce forest and fire laws and impress citizens to aid during emergencies. The act further allowed counties to organize "fire districts" (at their expense). It granted to the forester responsibility for the state's park, Big Basin. And it permitted the state forester, in times of "particular fire danger," to staff fire patrols with the cost borne by counties. In effect, California did on a state level what was happening on the national scene. Appropriately, the state's first hire was E. T. Allen, from the U.S. Forest Service.

The state, in practice, contributed little. The counties paid for patrols, the fire wardens were volunteers, and the only effective fire protection force was that furnished by the newly endowed USFS, supplemented in Southern California by organizations committed to protecting watersheds. In 1911 the Weeks Act allowed for formal cooperation and grants in aid between the federal government and qualifying states ("qualifying" meant the state had to contribute funds to the common cause of fire protection). Mostly, on-the-ground firefighting was the work of private range, timber, and watershed associations which contributed labor in kind and occasionally funds for trails and fuelbreaks.

The breakthrough came in 1919 with two new laws. One reconstituted (again) the Board of Forestry, which eventually came to be known informally as the Forest Department. The other permitted the state forester to create administrative units and appoint state fire rangers to supervise them, and granted some funds. The upshot was, in principle, an integration of government fire services. California could now join the Weeks program and claim federal subsidies; and it could operate on rural lands in ways jointly financed with the counties. In 1923 a public outcry halted an 80 percent cut of the forestry budget. Instead, the legislature enacted statutes under which counties could create fire protection districts and forest landowners could be charged for fire protection by the state if they did not maintain a "fire patrol" on their own. Few counties outside of Southern California took advantage of the act. The upshot was to increase state responsibility.

Still, an actual presence was meager. The 1919 program consisted of four patrolmen hired seasonally. In 1922 the Forest Department erected its first fire lookout. By 1923 the state had 16 rangers, four inspectors, and two lookouts. The next year Congress upgraded cooperative fire programs with the Clarke-McNary Act, which quadrupled the federal contribution. Cooperative agreements with national forests allowed for mutual aid along shared boundaries. Twenty counties contracted with the state to provide some level of protection. But field results were still lean. In 1927 the Forest Department had a staff of 28 rangers, six patrolmen, seven inspectors, and nine lookouts.

The reality was, the state never appropriated enough to do the task it set for itself. Instead, it looked up to the Feds for grants and down to the counties, fire districts, and landowner fees to staff for protection—and to a state emergency fund, whose expenditures fluctuated from $50,000 to $300,000 annually, "a huge sum compared with the regular State Forester's budget of those days," as Raymond Clar observed dryly. The state had effectively miniaturized the national system, with itself assuming the role of the federal government disbursing to the counties as the USFS did to the states and relying on emergency supplements to cover shortfalls. Increasingly, the cost of big fires forced it to consider a quasi-permanent staff of patrolmen instead of pickup labor. The apparent logic of fire argued, to many minds, for the state to create its own fire service for rural lands rather than outsource the job to others. The evident logic of politics and finances argued, however, that such a conception was a chimera.[3]

The Great Depression changed the calculus. Again, as though an echo of the federal Forest Service, California's sponsored a comprehensive survey of the issue (the Sanford Plan as surrogate for the Copeland Report), found additional funds and staffing, and built out an infrastructure in short order. Emergency monies and the Civilian Conservation Corps helped match means with ends. California became the largest and most audacious arena for CCC presuppression projects, of which the 650-mile-long fuelbreak known as the Ponderosa Way may claim special honors for hubris. Between the New Deal's Works Progress Administration and the CCC, emergency programs erected over 300 lookout towers, 9,000 miles of telephone line, 1.16 million miles of roads and trails, and numberless fire stations, tool sheds for smokechasers, and office and storage buildings.

With local options flattened by the crisis the Department of Forestry stepped in to make the case for a California-wide "master fire plan." Its essence was the belief that the various jurisdictions of governance should each be responsible for their own lands, which would leave Forestry to assume fire protection on the remaining landscapes as designated by the state. The plan would do for fire what parallel schemes would do for water. No longer would Forestry concentrate only on sites of high-value timberlands and watersheds. As the only fire department capable of reaching much of rural California, it would extend its mantle over the countryside. It could provide consistent, measured protection—the only governmental entity equipped for the task. The legislature was ambivalent, appreciating the value of the service but alarmed about funding it. The general fund was in deficit. The CCC was being decommissioned. Repeal of the Compulsory Patrol Act of 1923 was imminent. For two years floods not fires had submerged requests for more revenue.

What spared collapse was catastrophe. The threat of war led to a scheme for civil defense, prompted by the War Department and the State Council of Defense—what became the California Fire Disaster Plan.

As an existing network already responding to emergencies, the California Department of Forestry (CDF) was enrolled and designated as a statewide dispatch system. The scheme was entirely rational, and almost wholly hypothetical. For over two decades California had only begrudgingly built up its rural fire capabilities, preferring to decentralize and cooperate in ways that left the state more a broker than a player. A lot of shoving went on during legislative scrimmages, but neither side could effectively move the other.

Then came Pearl Harbor, and written plans doffed hardhats and staffed engines and lookouts in the expectation of attacks or sabotage on the mainland. At this time Forestry was under the administrative aegis of the Department of Natural Resources. Its director, Kenneth Fulton, proposed a dramatic escalation in state expenditures in anticipation of the extraordinary effort the war would demand. It was a sum calculated to jar the most jaded observer—"considerably more than the total war-caused needs requested by 22 other State departments." In effect, CDF would become California's department of defense. The money came through. The master fire plan gave Forestry its marching orders.

Over the course of the Second World War, California completed the infrastructure—"all the essential features of a full blown ideal plan," as Clar remembered—that the CCC had begun. The state would furnish fire protection wherever its resources allowed. It would step in, if reimbursed, to augment county forces. It would finance emergency firefighting. The funds bestowed—money almost coerced—by the threat of war would become the new budgetary floor. The military added manpower to firelines. Conservation camps of both juvenile and adult inmates replaced the CCC, and then persisted as a permanent source of labor. In place of coordinating volunteers, CDF was on its way to becoming one of the dominant institutions for fire protection in the nation.

Of course problems persisted. Squabbling was incessant over who should pay how much and for what. What exactly were the expectations on state responsibility areas? How much should the state pay counties for contract services if they so elected? (There were five, four of them in the south.) How much should CDF evolve from strictly forestry issues into an all-hazard fire service? Always, too, even in flush times like the 1960s, the state failed to appropriate funds sufficient to what standards demanded. Instead, the funding gaps were made up through expenditures from an emergency fund, and if the state's account was in overdraft, through federal aid. A wartime emergency had created, almost overnight, a comprehensive system. A continual cold war on fire kept it running.

What makes the California scene distinctive, however, is not that a state agency swelled so large but that wildland and urban fire melded. Even as the state's landmass was being pulled toward one or the other pole at the expense of a rural middle, something caused those extremes to fuse, and to hold together for a common cause. Some factor had to act as a flywheel to keep the pistons that powered fire protection, whether in cities, parks, woodlands, or pastures, in sync. That is the historic role of the Office of Emergency Services (OES), and as its name suggests, what forced the disparate parts into a single engine of response was crisis.

No entity could cope with the scale of California calamity on its own; no one could keep constantly on hand all the materiel and personnel that an emergency might demand, and even planning for an "average worst" event ignored the nonlinear—very unaverage way—in which California cataclysms collided with California society every decade or so. When the winds shrieked and the flames poured over the ridges, neighbor had to help neighbor, and when that failed, the region reached further, as it did with water. But as much as stockpiled apparatus, an agency charged with responding needed communications adequate to ensure that requested help could talk to the requester, and that once on the scene the differently uniformed and equipped responders could speak to one another. It needed incentives adequate to break down tribal allegiances and protocols. It needed a profound external shock.

In 1941 the California legislature enacted a war powers act that bestowed on the governor authority over all civilian protection agencies, notably fire departments, in the event of attack or a declaration of war. The governor assigned that particular responsibility to the attorney general, who established a State Fire Advisory Committee to oversee fire protection across 10 civil defense regions. The group included representatives from the U.S. Forest Service, National Park Service, California Department of Forestry, state fire marshal, and the chiefs of the three largest municipal departments (Los Angeles, San Francisco, and San

Diego). The state forester chaired the committee. The provisions moved from theory to practice shortly after Pearl Harbor. War against Imperial Japan and Nazi Germany made possible a collective fight against fire.

When the war ended, the military crisis was discharged into civilian life. In 1945 the California Disaster Act replaced the Advisory Committee with a similarly membered California State Disaster Council. Probably earthquakes could have replaced the threat of invasion, but major tremblors came too rarely and randomly. A state of continual emergency demanded a cataclysm that would recur in place and time with some regularity. California fire, particularly Southern California conflagrations, was ideal. The United States, too, commenced what might seem a permanent state of war, first with Korea, then the Cold War, and a succession of regional hot wars. Each external threat—the onset of the Korean conflict coincided with the Soviet Union's first atomic bomb—boosted the capacity for internal reaction to hazards of all kinds. The state established an Office of Civil Defense directly responsible to the governor, which included a Fire and Rescue and Emergency Services Branch and resulted in a Fire Disaster Plan.

Over the next decade the program underwent almost annual upgrades, and no less significantly, it found pots of honey to offset the dose of threatened vinegar. In 1951, in a civil-defense version of the Weeks Act, the federal government announced a program of matching grants for state and local authorities to acquire fire and rescue materiel. The projected windfall was significant—a hundred triple-combination engines, 29 heavy-duty rescue trucks, 100,000 feet of quick-couple pipe, and the basics of a statewide radio system. Gaining access to this cache was a powerful incentive for fire services to sign on since the engines, when not called out for emergencies, would be housed in local departments. To ensure equity as well as efficiency, the hardware needed software, however, so the Fire Advisory Board adopted a protocol for distributing the Office of Civil Defense's largesse to its members. All would be available to all. The old practice of cooperative fire and mutual aid, previously restricted to shared boundaries of cities and forests, swelled to cover 100 million acres. As in other matters, California assumed the role and scale of a nation in itself.

The pressures mounted while the Los Angeles Basin filled out and tract homes shouldered against the ridge spines, debris dams, and interior hills even as conflagrations seemed to come as often as the summer drought: 1953, the Monrovia Peak fire; 1954, the Panorama Point fire; 1955, the Refugio fire; 1956, the Malibu and Inaja fires. Southern California resembled a pyric fault zone, with each stressed patch rupturing in sequence. Then came 1961 with the Basin and the Harlow fires along the Sierra foothills, and, most notoriously, the Bel Air-Brentwood disaster, which burned the backlots of Los Angeles city itself. Each appealed to OES for support and each in turn stimulated the demand for more.

The sprawl of fires appeared to outpace the blistering urban growth. Even the proudest, most autonomous fire department could not keep pace. The hits kept on coming. The Decker fire, the Loop fire, the Canyon fire—the blowups were killing firefighters as much as they were leveling houses and unsettling watersheds. But it was not enough to send more hose: the flames were crossing the borders that separated not only incompatible land uses but fire services. The Forest Service trained to fight free-burning wildfire; counties and cities trained to protect structures and evacuate civilians. It was not easy to reconcile those tasks, and for all its impressive dimensions, the institutional edifice was, up close, full of cracks and loose boards.

Still, this being California, only a truly Big One could rattle the still-lingering complacency and confidence. The jolt came in 1970.

That year big fires blew a thunderous rain of sparks through the gaps. From September 22 to October 4, 773 fires broke out, of which 32 escaped initial attack, blackened 580,000 acres, burned 722 houses and some 200 additional structures, and inspired the then-largest mobilization in California history. Of the 32 big fires, all but three were in Southern California, including the monstrous Laguna fire (160,000 acres) and a complex that swarmed over the celebrity Santa Monica Mountains. Under OES direction firefighting forces converged from across the state, and then from around the West. They came from the Feds: the U.S. Forest Service, the Bureau of Land Management, and the National Park Service. They came from the state: the California Division of Forestry, the National Guard, Conservation Corps camps, and OES's own reserves of engines and support. They came from local authorities: cities from San Diego to

Los Angeles to Oakland; counties from San Bernardino to Humboldt, and especially the "contract counties" of the South Coast—Los Angeles, Ventura, Kern, and Santa Barbara. Outside fire crews poured in, from Forest Service hotshots to the Southwest Forest Firefighters, Snake River Valley laborers, and local Hispanic field workers. Fire engines by the hundreds funneled south. Some 28 air tankers flew missions, and a flotilla of helicopters dumped water, retardant, and burnout flares. At a time when critics of the endless Vietnam War were arguing to "bring the war home," the fall of 1970 seemed to realize that ambition.[4]

The 1956 fires at Malibu and Inaja had advertised the impending crisis, the 1961 Bel Air-Brentwood fire had broadcast the message widely, and the lethal 1966 Loop fire had confirmed the high costs in money and lives. Each had yielded targeted reforms. But the 1970 fires sought to catalyze the whole, to imagine a collective response on a par with the state water plan or the reorganization of its university system. That charge fell to the Task Force on California's Wildland Fire Problem, which promptly and efficiently identified the usual suspects and prescribed the traditional cures. It recognized that the 1970 explosion had plenty of antecedents and would foreshadow many offspring if nothing substantial changed.[5]

The breakthrough came when Congress ordered another approach under the direction of the Riverside Forest Fire Lab. In 1971 a group headed by Richard Chase reincarnated the California fascination with systems engineering that dated back to duBois (this time updated by experiences from the aerospace industry) and sought to make the *process* of firefighting work better. The tangle of jurisdictions and jumbled hardware could be—had to be—made to function much more smoothly. While it seemed unlikely that planners could shake tract homes free of wooden shingles, zone out construction in the wind equivalent of mountain debris fans, or even agree on the ultimate purposes of fire management, it should be possible to improve firefighting. The outcome was Firescope.

Firescope (FIrefighting REsources of Southern California Organized for Potential Emergencies) had its research charter approved in 1973, moved into field trials in 1975, and went operational in a graduated series of expansions from 1977 to 1979. Ideally, it sought a systems approach by which data would flow in, predictions about fire behavior and counter-measures would be generated, and a collective response from resources pooled by many agencies would follow. In practice, the elaborate gathering of information and modeling of fire behavior fell by the way-side, and what remained was the essence of the firefight itself, a means by which to coordinate personnel and equipment from scores of agencies to a common crisis. By means of sophisticated software, Firescope sought to reconcile a jumble of hardware platforms and the organizational cultures that operated them. It would do for individual incidents what OES did for statewide crises.[6]

Almost immediately, however, the chasm between wildland and urban fire services threatened to undo the enterprise. They had evolved in utterly different ways: all they had in common were those moments when flame put them both at a common border. Researchers were astonished that the two cultures not only had different terms for apparatus and operators but struggled to find terminology both could agree to. The example all cited was what to call a machine that squirted water. Urban fire departments called it an "engine." Wildland fire departments called it a "pumper." The terms reflected more than different classification schemas; they were markers of different occupational cultures, such that discussions quickly pivoted on the heritage and relative strength of their experiences. Yet each lexicographical difference was sand in the gears of common operations. Each difference was multiplied by scores of jealous jurisdictions.

All politics being local, and border exurbs being the shared focus, the provincial city and county fire departments came to dominate. They insisted that all vehicles that pumped water be called "engines," that fire officers be chiefs, battalion chiefs, captains, or engineers, not fire bosses, line bosses, or fire management officers, that standards for apparatus and performance come from the urban rather than the wildland side. Firefighters would wear turnout gear; fire officers would display bugles of rank on their collars. In the end, cooperation meant co-option. And because what happened in one part of California had to reconcile with the rest if emergency callouts were to succeed, the triumph of the urban model in the south meant its propagation everywhere.

Still, it took another 15 years and more catastrophic fires to replace the centrifugal forces of the assorted institutions with the centripetal power of OES. Firescope and the OES coevolved. In 1971 OES updated the California Emergency Plan, which included its Fire and Rescue Mutual Aid Plan. The 1977 fire siege, during which Firescope was vigorously tested, was followed by another upgrade. In 1980, another big Southern

California fire year, which included the ravenous Panorama fire, OES assumed full management for the program, and the National Wildfire Coordinating Group examined the program for possible national use. In 1982 the incident command system was rewritten into the National Interagency Incident Management System, and the federal government ended its contributions. In keeping with tradition, some 60 percent of the original system remained unfunded.

The participating agencies appreciated that if they wished to realize the full opportunities proposed by cooperation, reforms would have to encompass the state and to embrace an all-hazard model. OES coordinated the effort to spread Firescope lessons northward under the auspices of the California Fire Information Resources Management System. The integration was completed in 1986, not only between north and south but between Firescope and OES Fire and Rescue Service Advisory Committee as well. The outcome was the California Emergency Management Agency and a commissioned needs assessment for the future. The reforms arrived just after the fire siege of 1985 and before the siege of 1987. Further stress tests on the system followed in 1991 with the East Bay Hills (Tunnel) fire and the 1992 Los Angeles riots. By allowing for instant cooperation when requested, the biggest fire departments in the country in effect got bigger.[7]

Firescope succeeded as few fire research projects ever have. Because the logic of cooperation and the catalyst of catastrophe demanded ever more, its operational core, the incident command system, went national and then international to underwrite a universal protocol for all-hazard emergency systems. Incident management teams went to Yellowstone in 1988. They went to the Twin Towers in 2001. They assisted in the recovery of the space shuttle *Columbia* in 2003. They joined the Hurricane Katrina response in 2005. Significantly, the National Incident Management System as a research and development program migrated from the federal land agencies that first sponsored it to the Federal Emergency Management Agency, which now oversees its further development. That shift, however, had already been anticipated in California through OES. It was a familiar California story of an intrastate solution that became a national norm.

In 1945, the same year it passed the California Disaster Act, the legislature enacted a Forest Practice Act that established a permit system for range burning, and appropriated funds to purchase lands for a state demonstration forest. Forestry had to respond to controlled burning, wildfire, and timber harvesting. In its conception the agency remained a predominantly rural presence. But big fires, urban sprawl, and an institutional conscription for cataclysms all pushed the California Department of Forestry away from its origins.

The postwar economy, after a boom decade, shifted from commodities to amenities. Cattle moved from ranches to farms and feedlots, forestry meant recreation and biodiversity, not timber harvesting, and wildfire bolted out of wildlands and the rural countryside for an urban fringe. Even with a third of its responsibility lands lightly populated, the counties (or local fire protection districts) that contracted with CDF for emergency services were either filling with houses and malls or had their institutional geography deformed by such developments. What happened everywhere in California happened with CDF: the urban model dominated. Meanwhile, even as demand increased, CDF's funding from traditional sources shrank, quickened by the tax revolt that culminated in Proposition 13 in 1978. Though it sought to update traditional practices with such measures as the Chaparral (later, Vegetation) Management Program, its change in context from land management to sprawl servicing inexorably changed CDF.

The one constant was fire control. Even rabid tax protesters demanded protection, and big fires, as emergencies, stood outside routine budgets. More and more its fire mission defined the agency. Its original uniform patches featured a circle with a green conifer in the center. In 1979 its patches balanced two parts, one with a green tree and the other with a red flame in a triangle. In 1987 the agency changed its title to the California Department of Forestry and Fire Protection. In 1999 it eliminated the title "ranger" in favor of "chief." Collar brass now identified ranking. The next year CDF abandoned its classic green and khaki uniform, long the trademarks of a forester, for the navy blue favored by urban fire services, particularly Los Angeles County Fire Department. In 2006 the agency completed its transformation by relabeling itself CalFire. It had become an urban fire service in the woods.[8]

Paradoxically, by narrowing its land management mission into emergency response, CalFire had grown large. By the time of their shared centennials in 2005, it was second only to the U.S. Forest Service nationally as a fire and emergency agency. It exercised primary responsibility for 31 million acres and provided a degree of emergency services for 36 counties. It had a permanent staff of 3,800, a seasonal boost of 1,400, and a conservation (inmate) corps of 4,300 arrayed into 39 camps. It owned 23 air tankers and 11 helicopters. It had 58 bulldozers and 38 aerial ladder trucks. It responded to over 5,700 fires annually—and more than 300,000 incidents. It had an operating budget of $775 million. How much it actually spent depended, as always, on big fires, busts, and sieges.[9]

The saying that "fire does not respect borders" is, like many truisms, a half-truth. Flames certainly ignore boundaries that do not, in fact, bound anything other than names, and the phrase comes with a tinge of scorn where borders signify political entities that attempt to divide on a map what nature holds in common. Yet borders can also join together what nature has sundered. Both trends characterize California.

The need to respond to overwhelming crises, particularly wildfire, forced agencies to cross lines. If an agency stayed only within its jurisdictional boundaries, it would fail. The scope of cataclysm would overwhelm it. Yet in devising ways to cross boundaries, California the state unified what otherwise did not have common cause. It cajoled, coerced, cooperated with, co-opted, and otherwise cojoined what nature had sundered. It bonded Sierra redwood with coastal sage, Los Angeles with the San Gabriels, San Francisco with the Ventana Wilderness, and that most fungible and intangible of all intrastate divides, Northern and Southern California. Northern California found it difficult to schedule prescribed fires since its crews could be sent south; fire seasons, north and south, were out of sync; but they were merged under the master fire plan dedicated to suppression. The impact of these reforms could transcend California altogether: an "incident" became a category dissociated from any particular land and its history. The ICS could consider Maine in the same breath as Texas.

Those forced fusions demanded a powerful jolt of energy, and one that could repeat itself. California found that essential catalyst in recurring catastrophes, genuine or imagined. In the case of fire, the cataclysms were all too real, all too frequent, and all-too-often prone to border crossings. In the end, even California could not contain them. As both its critics and partisans have long believed (and feared), California, it would seem, is unbounded.

CAJON PASS

PORTAL TO SOUTHERN CALIFORNIA

S AN GORGONIO MOUNTAIN, where the Peninsular Range pivots into the Transverse Range, stands as a sentinel to the Los Angeles Basin. To each side lies a major pass into Southern California. Between it and Mount San Jacinto, to the south, lies San Gorgonio Pass. Between it and the San Gabriel Mountains, westward, lies Cajon Pass. Here the Old Spanish Trail emptied into the basin. Cajon Pass was the historic portal that joined Southern California to the rest of the nation.

It remains so today. Each successor transport system has built on that early trail. The California Southern Railroad punched through to connect the Santa Fe Railroad to the basin and San Diego. Other utilities followed—17 in all. There are two more railways. State Highway 138 got a big sibling in 1969 with Interstate 15. Southern California Edison runs three high-voltage (500kV) power lines and two 237kV lines. There are four oil and natural gas pipelines and five fiber-optic cables. And for backpackers the Pacific Crest Trail runs over the summit. The I-zone refers here to infrastructure. Cajon Pass is a femoral artery to Greater Los Angeles. If something shut it down, the effects would ripple not only throughout the South Coast but the nation.

Yet Cajon Pass is also a natural corridor for wind. Cajon and San Gorgonio Passes are among the most routinely windy sites in the state. More significantly, they are prime portals for the seasonal Santa Anas that drive the most volatile conflagrations. This means fire will flow as much as rail traffic, natural gas, and digital bits. In this extreme half-built environment, however, fire cannot be tolerated. To close down the corridor for even a few hours would have effects that could cascade throughout the country for days. Cajon Pass is a corridor: it is equally a chokepoint. It must be kept open.

So it mandates fire protection, but of a peculiar sort. It requires a fire service that possesses the intensity of urban firefighting but can operate within a setting that most people would characterize as wild. Fire engine meets fire wind. High tech meets high geology. The Mormon Rocks station halfway through the pass has the second highest call volume in the national forest system. Sycamore Pass station at the bottom has the third largest. (The largest, Oak Flats, lies on the west side of the San Gabriels.) Cajon Pass not only symbolizes the fire challenge of Southern California but is itself a forcer of that system.

———————

The San Bernardino National Forest has long prided itself on a fire organization that can match its mountains. Fire is to the San Bernardino what timber is to the Olympic or recreation is to the Tahoe. Its on-forest resources—a fleet of engines, four hotshot and two Type II crews, a rappelling crew, helicopters, including two Type 1 helitankers, a dozer module, an air tanker base—are commensurate to that charge. It originated the idea of a Forest Service honor guard.

But the San Bernardino provides an infrastructure, and has a reach, well beyond its own borders. It hosts the Regional Interagency Wildland Fire Training Center. It oversees the San Bernardino Air Base, a regional facility converted from the decommissioned Norton Air Force Base— the largest and most modern in the country, and usable by air tankers and heavy helicopters (two of which are on contract), which it can fill with water, foam, or retardant. It operates a Federal Interagency Communications Center (FICC) that handles emergencies of all kinds, including law enforcement, across nearly 30 million acres. The center runs 24-7, 365 days a year—the busiest FICC in the nation. It is an urban all-hazard model adapted for fire in the Mojave Desert, the Transverse Range, and the edge city. All of this occurs within sight, literally, of the country's largest metropolis and amid the highest concentration of media in the world. It is an amphitheater in reverse: the fires burn as though projected onto a giant IMAX screen. The air base even has bleachers so the public can watch.

What drives the system is not just that the San Bernardino Mountains have fire, but that they have big fires that threaten massive assets and act out in full public display.

The firefight is the great set piece of the American way of fire. But Southern California has bulked it up with performance enhancers until it stands to the rest of the country as San Jacinto Mountain to the Salton Sea. It has slammed the big fire against a big built environment. Wildland fire has to deal with structures, and urban fire has to cope with brush and hills. As the space between flame and city shrank, the decision space for fire's management went up in smoke, and the distance between the fire traditions that had emerged for each realm disappeared until they fused into something distinctive to make one of America's informing fire cultures. Unlike Florida, it did not result from a new graft onto an old rootstock as fire practices devised for hunting and ranching adapted to new purposes. Unlike the Northern Rockies, it did not evolve by reincarnating old practices into new avatars, a novel way of living in expansive mountain wilds. Unlike the Great Plains it was not a seasonal ritual of working landscapes.

In Southern California fire management means fire suppression. It means pushing an urban fire service into the frontcountry of a mountainous backcountry. For firefighting, this is the big time, animated by an if-you-can-fight-fire-here-you-can-fight-it-anywhere attitude. Everything is magnified; the flames, the costs, the aftershocks. Between 1990 and 2010 some 85 percent of structures burned nationally in the intermix have been in California. Today, some 50 percent of the Forest Service's national fire budget goes to California, and over half of that to the four urban-flanking forests of Southern California. Public scrutiny gets broadcast through a media bullhorn. There is scant margin for error. There is little tolerance for failure. In Southern California the Big Ones not only leap over mountains. They can, through institutions, cross continents.

Cajon Pass is a portal to another world. Through it Southern California has drafted wealth, energy, and ideas from outside, even as it projects itself outward to the rest of the country. Stand at Cajon Pass and see what this means and how it happened.

FOUR FORESTS

Southern California's Fire Rectangle

THE SIGN OF THE FOUR:
MOUNTAINS AND METAPHYSICS

EVERY MAJOR FIRE REGION has its metaphysics. Each has some question that assumes the form of a dialectic that it debates with passionate intensity, every partisan claiming "the science" is on his side, yet with no resolution in sight.

In the Southeast it is the issue of seasonality of burning, which varies as the land is put to new purposes. In the Southwest it is a controversy over cattle and climate, whether the El Niño-Southern Oscillation or humans "drive" the fire regime, and hence whether people can intervene effectively or not, be it for extraction or restoration. In the Northern Rockies it focuses on torches and lightning as the source of historic ignitions, which is to say, the original, putatively wilderness character of the land. If all these controversies seem irresolvable, it is because, scientifically, they are. The real issue is not about how nature works but about how we should work with nature.

In Southern California the metaphysics swirls around fuel and wind—which factor accounts for the eruptive Big Ones? If fuel, then there is a case for modifying vegetation to abate the likelihood of conflagrations. If wind, then there is little one can do in the landscape, and attention should shift from the landscape-scale source of the threat to the objects threatened. The stakes are high. The science for each has, so far, convinced its partisans, but not its critics. The default position becomes to prevent ignition up front or if a fire starts to hit it with overwhelming force before it can spread.

That is the discourse of the research community. Yet if you talk to the people on the ground, the polarity fractures into a regional vernacular that identifies a four-part taxonomy of big fires. Fuel fires, wind fires, terrain fires, arson fires—together they remake the traditional fire behavior triangle into a fire rectangle, and in curious ways they align with the four national forests that delimit the fire province. Of course each forest—every fire—features all these traits, but each might also be seen as displaying one of them with a special accent. The Big Kink is as much a wrinkle in pyrolexicon, and perhaps in pyroconceptualizing, as it is in pyrogeography. The time has come to recast the issue. The four mountain forests suggest how this might happen.

FOUR RANGES, FOUR FORESTS

Begin at the Mexican border.

The Peninsular Ranges run northward, bounded by narrow coastal mesas to the west and deep depressions on the east. They are relatively low, their crest a tumbling plateau tilted down to the Pacific. This is the realm of the Cleveland National Forest. The sprawl that is San Diego lies between the summit and the sea. The city includes some 900 miles of ravines scratched through the mesas.

The mountains abruptly pivot westward at Mount San Jacinto and San Gorgonio Mountain, rising 10,800 and 11,500 feet respectively, with a pass between them at 2,600 feet. The range pushes west as the San Bernardino Mountains. The passes to each flank are the main corridors into the Los Angeles Basin; the mountains stand like sentinels, and the San Bernardino National Forest guards the guardians.

The westward tracking range is known as the Transverse. A dense massif, the San Gabriel Mountains, occupies its center and looms over Los Angeles. To each flank, east and west, are gaps between mountains that provide the main corridors northward, and the primary flumes for winds. This is the domain of the Angeles National Forest. Then the Transverse Range breaks down, as though it had risen too steeply and the top

tumbled toward the coast into a series of smaller ranges that form a cata-
ract from the high desert to the shoreline, ending with the Santa Monica
Mountains. No mammoth mountain, no national forest—the mélange
of hills and valleys, like a giant sluice box between massifs, is overseen
instead by the Santa Monica Mountains National Recreation Area.

Then the mountains pivot northwestward. As if to make up for its
slacking, the range thickens into a terrestrial triple junction. A splinter,
the Santa Ynez Mountains, continues west, while the bulk splits northward in two. One branch, the San Rafael Mountains, moves northwesterly to merge with the Sierra Madre along the coast and the other, the
Tehachapi Mountains, flexes northeastward to meld with the Sierra
Nevada. The Coast Ranges trending north of the Transverse are what the
Peninsular Ranges are to the south. They continue until the San Andreas
Fault strikes out to sea at San Francisco Bay. For over 200 miles these are
the responsibility of the Los Padres National Forest.

That is the geographic matrix. Their geopolitical setting displays an odd-
ity as distinctive as the Transverse. In 1891 a rider to the Sundry Civil
Appropriations Act allowed the president to create forest reserves from
the public domain by simple proclamation. The first exercise was to
establish a buffer of reserves around Yellowstone National Park. The second was to proclaim forest reserves in Southern California. In most of
the West a public outcry arose, loudest from the monied interests, that
protested the "locking up" of natural resources. The Southern California
scene displayed the opposite logic. It was the local boosters who worried that unregulated cutting, grazing, and burning on the mountains
would destroy the watershed upon which the future metropolis would
depend. They linked fire and water; they saw the national forests as a
public investment in infrastructure for both. Fire control was the flip side
to water development, as vital as reservoirs and canals. The core mission
of the national forests was to stop the fires that stripped away the crust of
chaparral that alone held soil and retarded flooding.

That notion reflected the scientific understanding of the day. Forests, even if they consisted of head-high shrubs, were most valued for
their ambient "influences." They stabilized watersheds, they moderated

climate, they buffered lands otherwise prone to binge cycles of drought and deluge. Clear-cutting on mountains destroyed that capability; so did the trampling of hooves and the tearing by teeth of sheep, goats, and cattle; and so did fires. In Southern California the typical fire was a slope-scouring burn that left ash, dry ravel, and a field of shrub stumps. If the subsequent rains were heavy, the hillside could slough off in debris flows. Large chunks of the foothills were in fact formerly mountain flanks that had washed to the basin. The climate was mediterranean, which meant not only an annual cycle of winter rains and summer drought, punctuated by roughly decadal rhythms of wet and dry, but of mediterraneanity, which is to say a realm in which averages mean little. What drives the system are the extreme events: the big earthquakes, the major floods, the conflagrations. What haunts administration is the specter of the Big One.

THE BIG ONE

The fear is not unique to Southern California. The big fire has long dominated the American fire scene. It burns the most combustibles, it does the fullest biological work, it hijacks the loudest media, it costs the most (by far), it drives policy, it commands the most acerbic political attention. The 10 a.m. policy was designed to stop the big burns by extinguishing all little ones. When critics objected, the invention of the prescribed natural fire was an attempt to grant big fires range to roam. Even a casual observer might well conclude that American fire policy is a reaction to big fires. But whether or not that observation applies to the country overall, it certainly holds for Southern California.

From the onset, then, administrators have sought to understand the causes behind the big fires, which would then allow them to intervene with maximum effect. But the science has faltered—proved inconclusive, or contradictory, or too patchy for practice. Like generals refighting the last war, or economists preparing for the last recession, fire scientists find themselves arguing over the last burn, even as the landscape is changing around them. They want to identify a causal driver, as though fires ran like a commuter train and identifying the engine is sufficient to explain where and when it moves. Instead, fires more resemble a Google driverless car that barrels down the road integrating everything around it

without any single hand on the wheel. The Southern California reality
resembles a freeway matrix in which drivers swarm over the landscape.

Like Los Angeles—the metropolis without a core—South Coast fires
mix and match their contributing causes. This acentric etiology matters
because the identification of causes guides what responses fire agencies
will take. The assumption of course is that science will determine those
causes and prescribe cures. The reality is that, in Southern California, science stumbles, and economics and metaphysics fill the void. The reason
is simple. The act of settlement changes how the pieces come together,
and where the frontier moves quickly, as it does in Southern California,
fires' setting mutates rapidly. The fire environment is not something apart
from people. It results from the interplay of geography and people. Even
acting on ideas about how fire works can change the setting under inves-
tigation, and has. The Southern California fire scene did not evolve in
methodical sequence, as in much of the United States, but in a great rush.
The social construction of fire has been as spasmotic as its geography.
This is mediterraneanity powered by internal combustion.

═════════════

Every rookie learns the fire behavior triangle and how fires spread
according to three factors—fuel, terrain, and weather. In Southern California each of these factors is so magnified that each by itself can, under
the right circumstances, override the others. The interplay between them
is like the child's game of rock-paper-scissors. Wind cuts fuel, fuel covers
terrain, terrain channels wind.

The vegetation is, in parts, ever combustible, and under the right circumstances explosive. It can burn against wind or downslope: it burns
as gasoline might burn regardless of whether it forms a pool or splashes
over the ground or wafts through the air as vapors. Terrain is so steep
and crenulated that fires can always race up or send spots across a narrow
ravine, and firefighters have, after several lethal lessons, learned not to
build line on steep slopes under flame fronts, or try to cut across narrow
ravines on mountain flanks. And as to weather—the climate is a diction-
ary definition of *fire-prone*, and the foehn winds are among the worst in
the world because they come when the rest of the fire environment is
maximized for burning. The winds can drive flames down slope at night

and loft fire over sparse fuels or fling embers across the eight-lane San Diego Freeway.

The fire behavior triangle, however, only describes a going fire. There is no accommodation for when and how it might start; what matters is how it spreads. In Southern California, however, when a fire starts determines whether it gets big. Ignition is as fundamental to the ontology of big fires as the other three factors. An "arson" fire is as distinctive a category as a "wind" fire. So powerful and efficient are the regional firefighting systems that most sparks are quickly swatted out. What a big fire needs is an ignition so cunningly timed that it catches the other factors at extreme values and so escapes initial attack. Typically, that means a wind-kindled accident such as a broken power line or outright arson. It means a spark that is stronger than the most muscular firefighting complex in the world.

Each forest exhibits all these features, but each might also stand for one distinctively and serve to illustrate the complexity of that single factor. Consider what follows as an introduction to the systematics of Southern California fire and an abbreviated tutorial on why the search for understanding that can lead to remedies is trickier than one might assume. Southern California no more fits outside models of fire management than Los Angeles does models of urban development.

FUELING THE SAN BERNARDINO

Fuel is shorthand for an absurdly complicated mélange of combustibles. Calling it all "chaparral," or brush, is a further simplification, since chaparral is a generic term that absorbs grasses, shrubs, trees; sage, buckwheat, chamise, manzanita, ceanothus, scrub oak; natives and exotics; broad swathes and intricate niches nestled by springs or in the crannies of a furiously crinkled terrain. Lower elevations differ from upper, north-facing from south-facing. On the Transverse Range the upper reaches are conifer forest, and at the summit, tundra. What appears homogeneous from afar breaks down into hundreds of species, scores of habitats, and numberless combinations, all with complicated histories.

Labeling it "chaparral" is like lumping the entire human settlement in the Los Angeles Basin as "LA." The South Coast embraces the indigenous and the alien, the old time pueblos and the nouveau riche gated community, the giant and the miniscule, the City of Los Angeles and the City of Industry. What seems like an unruffled lake of suburbs, upon closer inspection, shatters into a fine-tiled mosaic of ethnic enclaves, 85 political entities, and patchy communities, only loosely united under a collective economic climate.

To lump the mountain biota into a "fuel problem" is like saying LA has a crime problem. The dismissive phrase disguises more than it reveals. But a good place to begin parsing what fuel means in Southern California is the San Bernardino National Forest.[1]

The San Bernardino Mountains have all the usual flora and fauna. But uniquely among the Transverse Range they have an active history of human settlement along the summit. American landscaping began with Mormon settlers who in 1851 established the original town plats for San Bernardino. Searching for timber they constructed a road up Mill Creek, and then wormed it upwards to the summit; mills followed.

The crest widened into pocked terrain sufficiently wide to support communities, and over the next century a cavalcade of miners, loggers, herders, tourists, and rangers trekked up to the top, remaking the scene. Much of the big timber was cut out. The grasses and low shrubs were grazed away. Valleys were flooded. Hillsides were dynamited and hacked apart by picks. What wasn't eaten out or hauled off was burned. In 1893, responding to local demands, President Benjamin Harrison designated the unpatented lands as the San Bernardino Forest Reserve. Another dozen years passed before effective protection arrived, when the reserves were transferred to the U.S. Forest Service for administration. Meanwhile, on the private lands, small communities took root.

Over the years protection and development advanced like partners in a three-legged race. Hydroelectric dams and debris basins filled lower ravines. An infrastructure for tourism matured, highlighted by such amenities as the Big Bear Lake ski resort, the Arrowhead Springs Hotel, and the Rim of the World Scenic Byway. Fire control strengthened, especially during the 1930s, the heady days of the CCC, which erected lookouts, cut trails, built fire roads and fuelbreaks, and fought fire. At times, however, the two stumbled over each other. In November 1938 Santa Ana winds drove a wildfire from Twin Peaks through Arrowhead and leveled the hotel. Eventually, as sprawl lapped against the base of the mountains, communities there also suffered; the 1980 Panorama fire was the most spectacular, burning over 345 homes in a single evening gulp. But by now, and for some decades, the general terms of land use, and hence the mountain's fuel array, had been set. What changed were their internal dynamic and the borders they shared.

The pressures only built, never released. Like the tectonic stresses that kept lofting the mountains upward even as they eroded savagely, big fires occasionally ruptured the landscape, which then rebuilt and readied for the next break along this biotic fault line. At the onset of the new millennium, however, the strains had reached new intensities. The mountains wouldn't wash away, the winds wouldn't stop, and the fires wouldn't cease. The nature of settlement had changed from rural working landscapes to one shaped by postindustrial amenities, so fire causes had morphed accordingly, even if they were invisible, like seeds banked in the soil. The bottom line was, fire would not fade away. Under exceptional conditions, some starts would escape initial attack and threaten to undo all that formal protection otherwise offered.

That left those amorphous fuels as the one point of vulnerability, the single article in the mountain's fire constitution that a fire management agency might amend. But fuels included the dead debris lying on the ground, the living verdure filling every hillside nook, and the wooden-slabbed houses and malls that extended from asphalt to lakeside and dappled the summit like cords of firewood. Homeowners associations often prohibited cutting. Logging of any kind ceased by the mid-1990s. Old fuel-break systems decayed. Then an extraordinary drought tightened the vice. It parched woods too thickly stocked to slack their thirst. Beetles—western pine, red turpentine, engraver—swarmed over the weakened trees. The die-off was ferocious, afflicting roughly 360,000 acres in all, the kills as dense as 600 to 800 trees an acre in places. Oaks died. Chamise died. Even cheat-grass, the living dead of invasives, died. Fearful of broken lines that might kindle a fire, Southern California Edison shut down power whenever the winds rose over 30 miles per hour. At the onset of the new millennium all the bad boys that made for big fires were doped up and ready for trouble.

Threat became fact in 2003 with the Old and Grand Prix fires, the San Bernardino's contribution to the regional complex. The cause of the Old fire, which burned 91,000 acres across and even over the mountains, was listed as "unknown." The combustibles it burned were not.

———————

The 2003 season traumatized residents and galvanized their congressional representative, Jerry Lewis, who conveniently chaired the House Appropriations Committee and liberated funds under the National Fire Plan and, later, the Healthy Forests Restoration Act. But the private sector, alarmed over potential liability, contributed as well. Southern California Edison eventually spent nearly $200 million to clean up around power lines. And residents joined in what became a reverse NIMBY. No one wanted to reside next to a vat of kindling. What evolved was a fire cooperative, the Mountain Area Safety Taskforce (MAST).

MAST had money, political clout, and public support. This was fuel mitigation as large-scale public works. Homeowners and utilities took care of private holdings, although some public monies came from the Natural Resource Conservation Service and the Forest Service; a program, Forest Care, also with federal assistance, would support upkeep. Otherwise, officials identified for priority treatment such strategic sites as roads necessary for emergency egress and entry and communication complexes. Fuelbreaks widened highways, major forest roads, and boundaries where they abutted houses. Communities created buffers. The Forest Service spent nearly $100 million in fuels treatments. Then more fires burned in 2007. One, the Slide fire, took out an area still under National Environmental Policy Act review. (The San Bernardino has more threatened and endangered plants than any other national forest.) Another, the Grass Valley fire, sent Southern California Edison a hefty bill when one of its power lines broke and started it—this even after treatment. The program began to move into maintenance mode.

Almost everyone associated with MAST regards it as a success. It was a proof-of-concept project that demonstrated what a dedicated program required and could achieve. Yet part of that formula was crisis and trauma sufficient to break through the indifference and gridlock of quotidian life and politics. And its cost has been high: it did to the national fuels budget what Southern California does to the national suppression budget. Once again the nation found itself underwriting the critical infrastructure for a state that long projected itself as an economic powerhouse. What California self-regarded as a source, others saw as a sink.

ASCENDING THE ANGELES

The mountains form a granite berm that bounds the region: for fire purposes, Southern California is what lies west and south of the chain of ranges hoisted up along the San Andreas. Those ranges seem as permanent as Earth itself, and their dynamism only that of the rhythms of deep time, irrelevant to the quick tempo of wind and flame. But in fact, they are many and varied, they affect every feature of fire behavior, and they can of their own accord dominate even a large burn.

Each mountain has its own character. The Peninsulars resemble squat fault blocks, tilting gently to the sea, gouged into mesas and channels. The San Bernardino and San Jacinto Mountains pivot around giant peaks. The western Transverse resembles a wooden branch flexed enough to break it lengthwise, so that it splinters above Ojai and fans out from the west-trending Santa Ynez upward into the San Rafael and Sierra Madres. The San Gabriel Mountains, midway in the Transverse, are most distinctive by being the simplest. They stand as a colossal massif, rudely shaped like a football, separated by major passes east and west.[2]

Those textured terrains influence how fire behaves, braking or boosting other factors. They make the Santa Anas possible by damming and compressing desert air, then directing the overflow through mountain spillways and weirs. They deform climate by letting air rise and fall over the great berm and fracturing a regional norm into thousands of microclimates and a mosaic of local weathers. What is true on a north-facing slope will differ from a south-facing; alluvial fans from summits; wrinkled ravines from exposed ridges. Vegetation will thin and thicken accordingly, display different levels of fuel moisture content, and crumble arrays of combustibles into biotic niches. Fuelbreaks will work when they function like levees, bolstering the terrain, and fail if they cross the grain of terrain-sculpted fire flow. And landscape texture will influence where and how fires begin by influencing the placement of roads and trails.

By guiding the movement of people those routes affect both fire starting and fire suppressing. People will kindle ignitions along routes of travel—literal lines of fire. Those same roads, trails, and fuelbreaks will transport firefighters and become firelines. Terrain will define natural units of fire spread and containment, and it will dictate—or should— how fires might be attacked. Over and over, on a 20- or 30-year rhythm, like a cycle of earthly sunspots, the big fires reappear in the same places, burning in more or less the same way. The vegetation regrows, the winds return, the terrain-crafted basins that contain fires' behavior fill with flame. Overignition and exotic grasses might quicken the cycle so they burn more often, as in the San Gabriel canyons, but the larger terrain inscribes the pixels that determine how and where fire goes. Defy that geographic logic and you risk losing lives as well as firelines.

That last cautionary precept is the story repeated sickeningly over and over in fatality fires as crews find themselves on steep slopes with fire below them. That lethal flame might be a reckless burnout. Or a hooking fire, creeping down from one ravine and entering another below a spur. Or a spot fire, from who knows where, that plants itself below a chimney and flashes up through anyone unwise or unlucky enough to be in its way.

It's doubly diabolical because the inclines that can shoot fire upward also slow firefighters to a crawl. Caught on a slope is what happened to a pickup crew at Griffith Park in 1933 and killed 25, to marines when the Hauser Creek fire killed 11 on November 2, 1943, and to a Forest Service inmate crew from the Viejas Honor Camp that killed 11 at Inaja on November 25, 1956. That is what happened to the El Cariso Hotshots on November 1, 1966, when they tried to cut line across and down a chimney on the San Gabriels and a spark sent a blowtorch of flame upward and killed 12. And it is what happened in August 23, 1968, when the Canyon fire on Glendora Ridge of the San Gabriels blew over eight firefighters from the LA County Fire Department (LACFD).

That burn vividly demonstrated the varied but cumulative power of terrain. Fire season was well advanced, but not notoriously droughty or prolonged. Winds were light—this was not classic Santa Ana weather. Fuels were mixed but generally also light—the fire burned through grass, oak, shrubs, and soft chaparral. Instead, what controlled virtually every aspect of fire spread was terrain. Repeatedly, fires burned across

Glendora Ridge ravine by ravine. Slowly it crept down, entered a new arroyo, and then rushed up. On the day of the disaster fire had "slopped over" the head of an adjacent ravine around 11 a.m. and painstakingly crawled down through litter below the mixed scrub. At approximately 11:24 a.m. it flashed in a patch of sumac and scrub oak. What made this flare-up similar to the 1959 Decker fire but different from others was that it captured a whirlwind—almost certainly the result of turbulence over the rough countryside—and became a fire whirl. The fire whirl hurled a firebrand below the ravine in which the crew from Camp 4-4 was working. Probably "not more than a minute" later, and perhaps no more than 30 seconds, that flash ignition had blown up, blasted through the ravine, and savaged the crew. The fire's behavior, a review board concluded, was "almost entirely controlled by the topography."[3]

That tragedy was keenly on the minds of the crews from LACFD and the USFS when, on August 26, 2009, 41 years later almost to the day after Decker, a fire broke out on the Angeles National Forest along the Angeles Crest Highway. Again, winds were light and fuels were volatile but not outside norms for age and season. Suppression resources were abundant. The critical circumstance for what became the Station fire was terrain.

It magnified ignition and compromised suppression. An arsonist had craftily kindled the start where it would do the most damage. The highway, the anchor point for a fireline, was midslope, which ranged between 33 and 67 percent. Visibility was poor, access feeble, foot travel tricky, and the opportunities high for spotting both above and below the point of origin. A board of inquiry concluded that the landscape was "extremely difficult, if not impossible, to traverse without a high degree of exposure to hazard." It effectively neutralized an impressive initial attack force: 9 handcrews, 13 engines, 3 water tenders, 4 medium helicopters, 1 heavy helicopter, 3 heavy air tankers, and an Airco helicopter, 2 patrols, and 4 chief officers from the 2 agencies. The setting simply didn't allow for such massive deployment, like a field army forced to fight in a city of narrow streets and alleys. Spot fires below the highway remained inextinguishable by aerial attack alone and yet inaccessible to ground forces. The serial spotting, which continued through the evening as flare-ups made runs upslope and flung firebrands, overwhelmed any chance at control. By the next afternoon a Type I incident management team had been ordered.[4]

The Station fire burned and burned. It burned day after day, week after week, basin after basin. It burned the grasses and soft chaparral at the bottom of ravines and the forests at the mountain peaks. It burned from its own momentum and from burnout operations along ridgelines. It billowed smoke behind Los Angeles like Mount Vesuvius outside Naples. It burned until October 10, 2009. It ended, that is, before the season for Santa Anas. At a final tally of 160,577 acres the Station fire was the largest in the annals of the Angeles National Forest and the 10th largest in modern California history.

Something this vast and this visible can be parsed many ways. It was an arson fire, of course. It was a fuel fire—there was more than enough to burn in all directions. But the ignition took because it couldn't be suppressed safely; the fuels were in no way uniform, unusually dry, or exceptionally heavy, and burned because slopes and narrow canyons channeled flames and scattered spots. The winds that propelled it were largely local, the outcome of diurnal warming and cooling in textured hillsides that guided flow up and down. The fire burned, in brief, because of where it occurred. It was a terrain fire. And it challenges assumptions about the taxonomy of fire in Southern California as much as it did initial attack.

———————————

But there is more. A fire's size does not measure its significance. A 100-acre fire in the frontcountry can trump a 100,000-acre fire beyond the ridgeline. What matters is how visible the burn is to the culture; whether houses burn, whether people die, whether someone is present to transcribe the smoke into pictures and words. Little can be hidden on the Angeles. It forms a backdrop to the largest metropolis in the United States and to a major media mecca. It is a backlot to Hollywood. As powerful, a state statute known as Rule 409.5, unique to California, allows all credentialed agents of the media full access to any emergency scene and makes the fourth estate a fourth side to the regional fire triangle. No one should wonder that the Angeles has the largest fire budget in the national forest system and co-manages with the largest-budget county fire department in the country.[5]

In some regions, geology obscures. It removes fire from direct contact with society, it renders the flames invisible. In Southern California the reverse holds. The San Gabriels thrust fire against the city with almost no mediating landscape, or media baffles, between them. A fire like the Station or the Gap or the Woodwardia can't remain out of public sight. It is impossible for an agency to perform without an audience of millions of critics and catcalls. In the case of the Angeles—here synecdoche for all of Southern California—terrain affects not just fire behavior but the behavior of people toward fire.

IGNITING THE CLEVELAND

The Cleveland is best known for its epic wind-driven fires—the 1970 Laguna (175,425 acres), the 2003 Cedar (273,246), the 2007 Witch (197,990). So monstrous were the burns that they overwhelmed public memory of fires that would, anywhere else, have been considered mammoth, such as the 2003 Paradise (56,700) and Otay (45,971) burns, and the 2007 Harris (90,000) and Poomacha (49,000) fires. What has shaped the narrative for them all is the interplay of diabolically dense chaparral and vicious winds.[6]

But Santa Anas and fuel could act only if a spark was present; otherwise they were simply wind and brush. The fires did not originate spontaneously, as if by manzanita rubbing briskly in the wind. For a fire to get large it had to happen with exactly the right timing: a point of ignition is, literally, the spark that transforms inert substances of earth, air, and biomass into a chemical reaction. The timing of ignition decides whether the other factors that shape fire behavior remain as mountains, ravines, airsheds, woods and shrub fields, or whether they become terrain, weather, and fuel. What might be characterized as wind-driven when it's moving might be equally typed as ignition-enabled for existing at all. In Southern California spark remakes the fire behavior triangle into a rectangle.

The sources for the big burns display a curious cross section of regional life in the new millennium. Overwhelmingly, the cause is people—their machines, their demands for electricity, their eccentric mastery over devices designed to replace open flame but which might spark it outside their wrapping, their clumsiness and malice. The Laguna and Witch fires started from sparks cast by a power line; the Cedar, by a hunter's signal fire; the Poomacha by a house fire; the Otay and Harris, from border crossers; the Paradise, from arson.

In brief, what is called ignition is no simpler than fuel. It embraces a Pandora's box of sparks that, when opened in the right conditions, fly out across the landscape like an ember shower. The sides of the rectangle affect one another. Spark is just heat until it connects with fuel and is fanned by wind, but once having kindled a fire it changes the future of the fire environment.

Some traditional burning has endured in the ranches and farms that dapple the backcountry. But mostly, the pyrogeography of ignition has been remade as fully as the landscape that receives it.

A typical profile of ignitions for a Southern California national forest looks like this: Instead of landclearing fires by the open burning of windrows, fires start from dozers and chainsaws and other equipment (37 percent). Instead of range improvement burns, vehicles (2 percent) and power lines (2 percent) kindle blazes. Instead of debris burns, the ignition exotica of modern American life, lumped together as "miscellaneous," call fire into being (14 percent). A few crossover categories persist. Smokers account for 5 percent, campfires for 4 percent, and that old bugbear, children with matches, 5 percent. Lightning, on the mountainous outer rims, claims perhaps 2 percent, roughly the same as car crashes. Some 8 percent start from arson. The giant La Brea fire began from a marijuana grower's propane burner; the Witch fire, from power line failure; the Montecito Tea fire, from students at Santa Barbara City College who failed to properly extinguish a bonfire. The Simi fire kindled from a long-range firebrand hurled from the Verdale fire. Strip firing by suppression crews can serve as new sources, interacting dynamically with the main fire to warp behavior. The source of a whopping 19 percent of ignitions is unknown.

While people decreasingly cook and heat houses and work the land with open flame, their hidden combustion more than compensates as a source of sparks. Moreover, the scene bears a deeper imprint, a palimpsest of past ignitions that continues to affect what fuels are available and how later fires burn.

Whether a spark on the ground leads to a fire depends on whether it can generate enough power to sustain itself, whether it can give off more heat than it absorbs. Most can't. In its early stage the heat from a kindling flickers insecurely between source and sink. That logic applies equally to the competition between fire starters and fire suppressors, or the capacity to suppress a fresh spark before it can propagate.

Once again, what appears simple fragments into the complicated. Viewed as a whole, Southern California is a fortress of firepower; of the thousands of ignitions that do take, only 2 to 3 percent get big, and of those perhaps 2 to 3 percent truly register in the regional chronicle. Around the Cleveland the two dominant presences are the U.S. Forest Service and CalFire. But the Bureau of Land Management is also on the scene; there are scores of cities; Camp Pendleton bristles with firepower; and there are 29 tribes with postage-stamp-sized reservations, some lit up with casinos, some vacated. (With 104 tribes California has the largest mix under Bureau of Indian Affairs jurisdiction.) It is a bizarre variant of the wildland-urban interface—a wildland-reservation (or wildland-casino) interface. Reservations with major casinos, as a condition of their compact with the state, must create fire departments, and some have wildland fire capabilities, although most contract with CalFire through the Bureau of Indian Affairs. That few fires break free is an astonishing tribute to the capacity of regional fire agencies to concentrate and cooperate.

Yet the crucial story may be one lost to modern sensibilities. What historically improved the odds for big burns was the fact that, in the past, every fire that took might linger on the land for days, or weeks, or even months. The critical timing that allows a spark to mate with high wind or thickets of fuel rises with each creep into a new ravine or simmering below a potential Santa Ana. Deal enough hands and you will win at solitaire. When crews lose a fire now, they do so quickly, under the improbable, but reoccurring, circumstances that instantly align the conditions that favor a start with those that favor roiling spread. Once the fire has made a run, forces contain it to prevent another. In the past, they didn't.

So in the reckoning about ignition, the capacity to stop is almost as vital as the ability to start. Probably the full number of ignitions has altered little over the years, if not centuries. What has changed is the capacity of the land to receive them and the capabilities of fire agencies to swat them out. Even under extreme conditions it usually takes multiple starts to overwhelm the system.

In the discourse about ignition one theme, for the public, trumps all others. That theme rides the saddle of a fear that masks a rationalization. The theme is arson.

Behind it lies an unease about the absence of organic community, that there is no intrinsic or internalized social control over behavior, only what can be imposed by force before the fact by the police or after the fact by the fire department. The recurring vision of Los Angeles is not only that it burns but as Nathaniel West portrayed in *Day of the Locust* that it burns from its own hands. The torch-wielding rioter, the pyromaniac, the arsonist—these rival serial killers in the gallery of public villains. And not without some truth. The Esperanza fire that killed five firefighters was set by a serial arsonist (later convicted of murder and sentenced to death). Of the dozen major fires that constituted the Fire Siege of 2007 two were started by arson and six had unknown causes, of which it is possible to suspect that at least some came from arsonists wily enough to hide their misdeeds. In some places arson is a nuisance, burning vacant lots and empty buildings; in others, an insurance scam, or a pathetic appeal for recognition. In Southern California it is, if done craftily, a lethal threat to the very texture of the built environment.

But an obsession with arson is also a study in political misdirection. It is easy to dismiss intractable problems by blaming the wayward stranger, the border crosser who lights up a campfire and lets it escape, the madman who responds to some inner compunction, or the impulsive incendiarist who, like a renegade cop, turns his knowledge to evil. All deflect attention away from the circumstances that give the arsonist his power, for a fire more resembles a riot than a drive-by shooting. It spreads. It derives its power from the power to propagate. What makes arson dangerous is that the setting in which it occurs is prone to burn. In Southern California that setting is not simply the chaparraled mountains but cities ill sited and houses ill designed that make fire a threat. The arsonist can increase the odds of setting a conflagration. But even the worst wind or densest brush cannot carry fire through a suburb designed to withstand it. The arson problem is really a problem with how Southern Californians choose to live on the land.

It is a structural failure, and its ideal illustration is the broken or arcing power line that casts sparks on whatever lies beneath it. That such events occur particularly during Santa Ana winds makes the mountain high-voltage line a significant source of big fires. In the 2007 Siege, across the region, arsonists are known to have started two blazes. Power lines kindled four. According to CalFire statistics, throughout the state power lines account for only 3 percent of fires (about the same as escaped campfires); but these ignitions almost always occur under conditions that favor conflagrations.

What would seem to be a fuelbreak, the swath cut by high-voltage transmission routes, can paradoxically be a long fuse. The utility companies might stand accused of structural arson, which is why they are willing to cut power during high winds or spend obscene gobs of money to clean up around their rights-of-way to avoid liability lawsuits. So, too, suburbs that ought to break a spreading fire can carry it if poorly crafted. It has taken 20 years of repeated firestorms to finally break resistance to suitable fire codes, even such no-brainers as abolishing shake-shingle roofs. Yet the rules apply only to new construction. As fires take out structures, they will be rebuilt to resist not so much the flames as the wind-wafted embers.

That process—hardening houses—might be likened to a public health campaign for vaccination. The more houses get inoculated, the less likely the spread. The way to best control arson is to take away the power of the torch, to keep an ignition from boiling over or if it kindles a fire that races with the wind to keep it from combusting houses like an emergent plague killing off a virgin-soil population. Probably it is too much to expect that, in such a celebrity-saturated landscape, arson would not become sought after and sensational. But the problem will only truly become domesticated when the landscape itself is tamed. At that point, arsonists, like rabble-rousing firebrands on a street corner, can declaim passionately day after day but the passersby won't pick up and spread the sedition.

BLOWING IN THE LOS PADRES WIND

Everyone knows that wind fans fires, and almost everyone knows that the winds that matter are the howlers, which in Southern California means the Santa Anas. But a finer-grained scrutiny shows the same complexity for winds that characterizes terrain, fuels, and ignitions. In the South Coast big fires can occur without big winds. The winds that matter are those that push fires against cities or that trap firefighters. Straddling a tangle of mountains, the Los Padres National Forest illustrates that distinction decisively.[7]

During a three-year span, from 2007 to 2009, the forest experienced five major fires. Two giants loped over the backcountry, pushed and pulled by landscapes rugged with ridges and lumpy with chaparral. Three smaller burns slammed into high-value frontcountry. The three were an order of magnitude (and in two instances, two orders of magnitude) smaller than the large but they took out houses in a celebrity-studded coastline. That happened because winds spilled over the mountains and rushed downslope, even at night. Two landscapes, two kinds of fire behavior, two relationships to wind, and two kinds of fire.

———

The Los Padres sprawls over that granitic briar patch where the Transverse Range breaks and splinters into a fan of lithic slivers—the Santa Ynez, the San Rafael, the Sierra Madre. It then, with great gaps, continues leisurely up the Coast Range to Big Sur. Among the four regional ranges it has the largest fraction committed to wilderness (80 percent). It stands closest to the sea. And it displays the fullest assortment of winds.

There are the local winds—the land and sea breezes that are part of daily rhythms, that rise and fall over the land as tides do for the ocean. Mostly they bring maritime air and frequent fog. Daily, too, on the backside of the mountains, winds strengthen and falter as the sun heats and the evening cools; they interact with terrain and clumps of fuel to send flames up chimneys, spin them through eddies and fire whirls, and quiet their spread at night. Overlying them are the synoptic winds that accompany fronts, or the rolling in and out of the region by high and low pressure cells. These, too, come from the west, or cause winds to veer from southwest to northwest, and they carry Pacific air to the crests of the mountains. Sometimes they supplement, sometimes they counter, the local breezes. This is all standard fire behavior with a California accent.

What makes robust fires into monsters are the winds that reverse this pattern and blow from the desert to the sea and from the crests of mountains to their bases. Where sea winds moisten, these desiccate; where slope winds rise and fall with the sun, these burn through the night; where normal fire can be flanked on ridges, these can spill over the levees and scatter embers with the abandon of a girandola. Such winds are not unique to Southern California. They belong to a category known, unimaginatively, as strong mountain downslope winds. They are the foehns of the Swiss Alps, the chinooks of the Rockies' front range, the east winds of northern California and the Pacific Northwest. But nowhere else do they interact with such ferocity or last so long as at the South Coast.

The conditions that make them possible are hardwired into the regional terrain and the software of climate. Every autumn high pressure can stall over the interior basins of the West—on smaller scales, the Great Valley of Central California, or the Santa Ynez Valley—while low pressure approaches from the southwest. The mountains stymie the flow between them, and where the mountains are deep and high, and the pressure difference between centers is great, the gradient effectively steepens, and when the winds break over the summit or pour through passes, they gush explosively. They flow over the mountains, dry, warm, gusting with the untrammeled sprawl of an avalanche or the floodwaters from a ruptured dam. In the Santa Anas the California tendency to exaggerate natural phenomena finds a meteorological expression. If a spark is anywhere around, flames take on the character of those winds. More insidiously, high winds can cause power lines to break or arc and scatter sparks, thus adding ignition to their gallery of malevolence.

But the winds, even under Santa Anas, are not quite so simple. They are channeled by passes through the mountains, broken and funneled by ridges and ravines, diverted by summit spillways. And they interact with local winds: the bellows that bring prevailing winds from the west and the local generators lodged deep in mountain valleys continue, and the Santa Anas must beat against them, skip over them, or blow them locally away. One can weaken and the other strengthen—how they engage will blow perimeter flames like leaves in a tempest and shape the contours of the burn overall. Still, there is a logic imprinted on the landscape, much as with floodwaters. There are patterns, although as with all matters Californian, from time to time the norm is meaningless, even when it involves the infrequent. What matters is the rare and the extreme.

To all this the Los Padres offers a local codicil. The Santa Ynez Mountains are a cameo of the Transverse Range, the coastal-front mountains for a granitic wave train. A deep valley behind, steep mountains against an urban complex—this is the South Coast formula in miniature. What the Santa Ynez Mountains are to the Transverse Range, the Sundowner is to the Santa Ana.

By rushing down to the Santa Barbara coast at dusk the Sundowner winds seem to embody their name. The etymology is otherwise, yet uncertain. In one version it derives from the Spanish *zonda* (for foehn wind), while in another, from *simoom*, a version of *scirocco* (perhaps via Spanish). Or those may be attempts by the colonizing Spanish to render the Chumash term into something that sounds familiar. The one surety is that English speakers converted the indigenous expression into something else that loosely related to the phenomenon as they experienced it.

The same seems to be true for its meteorological mechanics. In a loose way the Sundowner is a Santa Barbara Santa Ana. It relies on a similar arrangement of air masses that causes winds to pile up against the north slope and then spill over the south toward the coast. As they rush down, they warm and dry. It may be that the Sundowner is what remains after air cascades from the north across the fan of mountain waves. Or it may be the outcome of a mesoscale equivalent in which the California interior accretes air and a Catalina eddy tugs them south. In some cases the Sundowner effect means a dramatic rise in temperature (up to 107°F). In others, it means a downslope wind particularly noticeable at night.

The history of Santa Barbara is a chronicle of Sundowner fires. Sediment from the channel holds charcoal through most of the Pleistocene. The American record begins when Richard Henry Dana recounted one such episode that dated from the 1820s. In August 1940 the San Marcos fire "provided a thrilling spectacle for the thousands of Santa Barbarans and outside visitors celebrating the annual Spanish Days festival." Over the past few decades the fires have become less an entertainment and more a threat. The Refugio (1955), the Coyote (1964), the Wellman (1966), the Romero (1971), the Sycamore Canyon (1977), one after another, they burned swathes down the slopes of the Santa Ynez, rupturing along a biotic fault line, and all raised the stakes since the flames no longer passed through the rabbit habitat favored by the Chumash or the ranch-lands preferred by Spanish and early Americans. On June 27, 1990, a Sundowner drove the Painted Cave fire through 4,900 acres, 427 homes, 221 apartments, 15 businesses, and 10 public buildings. In July 2008 the Gap fire washed down toward Goleta and Isla Vista, half of its 9,544 acres on the Los Padres, half on private lands, upending watersheds. In November 2008 the Montecito Tea fire blew over 1,940 acres and 210 houses, Westmont College, and the Mount Calvary Retreat House and Monastery. In May 2009 the Jesusita fire rushed down Mission Canyon toward downtown Santa Barbara, taking out 8,733 acres, 80 houses, and part of the Santa Barbara Botanic Garden.

The right wind on a smartly placed spark will spread. So, too, even a small fire, measured by geographic size, may spread widely through the culture if fanned by modern media. Such propagation requires as primary fuel not chaparral but celebrities. Actor Christopher Lloyd lost a home. Rob Lowe had to evacuate. Steven Spielberg evaded the flames. Oprah Winfrey was spared but broadcast her feelings on her show. Governor Arnold Schwarzenegger visited the burn.

Fire effects, that is, are no longer limited to air, soil, water, flora, and fauna. With the right wind behind it a fire can blow ash across national TV. The Sundowner can not only leverage an ember into a conflagration but blow up a local fire into a national story. It is not the vigor of the wind that matters. It is its ability to shake the right neighborhoods and, through its funneling down media channels, to be made visible.

THREE PARKS

Traversing the Transverse

THE TRANSVERSE RANGE is long as well as high. Cut it in cross section and it exhibits one suite of fire regimes and management styles. Cut it lengthwise, however, and it suggests other ways the regional fire regimes are organized and might be administered. The cross sections are characteristic of the national forests; each replicates the other, with some local spices stirred into the common stew. The longitudinal traverse is typical of the region's national parks, each of which differs mightily from the others.

Geographically, the national forests divide desert from shore, and urban from wild. A cross section would go from high desert or Great Valley over the hump of the range, rising into rich woods, and then plunge down the other side into chaparral, grasslands, and suburbs. It would pass through fringe settlements, multiple-use forest, wilderness, and city. A thin slice passes through, as it were, the life zones of the region's fire management. One administration must deal continuously, seamlessly with the whole gamut, just as one fire might spread from bottom to top, or top to bottom. Fires burn, so to speak, with the grain of the country.

The national parks offer a different perspective. They run east to west along the arc of the Range. To the east, Joshua Tree National Park sits atop the Little San Bernardino Mountains, grading into the famous high desert that rises behind the Transverse Range as it makes its great kink westward. Santa Monica Mountains National Recreation Area (SMMNRA), a tumbled sequence of hills and valleys that replaces the San Gabriels to the west, spans from interior to seacoast. Channel Islands National Park is an archipelago of five islands, separated from California by a rising sea, but otherwise identical in rock and biota to the mainland. They are literal isles, small worlds unto themselves—chips from the Transverse block.

To traverse across the three parks is to sample the fire options of Southern California. They assemble the pieces of the regional fire scene into distinctive regimes, from desert to mountain chaparral to arid maritime; and although administered by a common agency, they have devised dramatically different programs. In simplistic terms, Joshua Tree has a problem with its flora, particularly the competition between pyrophytic invasives and indigenous species; the park is sprawling but coherent, though it bleeds into the world beyond. Channel Islands has a problem with its fauna, particularly the competition between invasives that strip out fire and an indigenous habitat that needs it; the park is sharply delimited but scattered, and each isle is isolated from its surroundings. Santa Monica Mountains, by offering a pastiche of a biota and a settlement pattern, has a problem determining what its true needs are amid a tangle of research studies that stumble over one another, and if it can decide what to do, how to implement a coherent vision amid its jumble of jurisdictions.

Collectively, the three parks show that many combinations are possible among the pieces that make up the regional fire scene. What, at first contact, would seem to be a coherent fire province fragments into a pile of child's blocks that can be assembled into many patterns. The parks accent the need for particularized solutions to particularized problems—and that is what the fragmentary, in some respects feudal, political organization of the national park system allows. It reconciles a larger vision and institutional capacity with the idiosyncrasies of specific places.[1]

CHEATING FIRE

Joshua Tree National Park straddles a mountain, but the Little San Bernardino (and its echoes, the Hexie, Eagle, and Coxcomb) run east of the Big Kink and so lie in the immense rainshadow of the Transverse. Two desert biotas converge, the Colorado to the south and the Mojave to the north. The Colorado is lower and drier, the Mojave more elevated and vaguely wetter. The mountains have juniper, blackbush scrub, and the fabulous Joshua Tree, an arboreal-looking relative of the lily found only in the Mojave, and the reason why the area was declared a national monument in 1936. The sparse vegetation, standing like runes in a graveyard, argued for a large (860,000 acre) reserve.[2]

Joshua Tree was then a forlorn and otherworldly place, heaps of rocks and strata like crustal scabs, lightly marked by vegetation, resembling a pointillist painting, visited mostly by prospectors. In 1950 pressure from mining interests succeeded in redefining the monument, removing 300,000 acres. Recreationists, including rock climbers, gradually discovered its peculiar charms. Travelers on the way to LA (or the Palm Springs casinos) made Joshua Tree a kind of motorized nature trail, easily accessible from I-10. Then the desert tortoise got listed. In 1994 the Desert Protection Act raised the monument to national park status and added 234,000 acres.

But if access brought public interest, it also brought ills. Like the Spanish missions that helplessly introduced disease and decimated indigenous populations, these new pilgrims carried exotic grasses, notably *Bromus*—red brome—and its better known cousin, cheatgrass. The grasses brought fire. A place that had known fire as it had earthquakes, continually but rarely, got more of it and more often.

The character of the natural regime is unknown. The desert is not gregarious with combustibles; its woody constituents stand apart, separated by desert pavement and boulders, and nothing connects the patches. A fire would flare and go out, like a flaming match head. Surely, in El Niño years, when even the great peaks could not stop rains pushing into the desert, the flush of moisture would cause the desert to bloom, and the blooms would then wilt into stalks, and sometimes burn. No record of scarred trees or charcoal varves chronicles such events; but it seems plausible. It's hard to imagine the place immune from fire. It's equally difficult to envision it routinely swept by flames. The species show few adaptations to fire, certainly none that indicate that fire might be necessary. The signature Joshua Tree recovers poorly and painstakingly.

Then the missing fuels arrived. From its roadside infestation the brome filled the porous biota like water flooding a landscape of shallow depressions and pockets. A fire could carry from patch to patch, and more insidiously, since the brome cured earlier than indigenous grasses and forbs, it could burn again, and again, in a positive feedback, and gradually drive out the indigenous flora. Meanwhile, along the north, spreading out from highways, military bases, and watering holes like Twentynine Palms, houses, strip malls, and other industrial tentacles twisted and squirmed through the passes and into the high desert. Together, like an invading virus, they commandeered the nucleus of the old fire regime.

Fire became possible, and then inevitable. Since 1965 fires propagated beyond their old quarter-acre allotments. In 1979 the Quail Mountain fire burned 6,000 acres. In 1995 the Covington fire scorched 5,158 acres. In 1999 the Juniper Complex blasted over 13,894 acres. The idea that modern fire management meant restoring fire became itself threatening. If the fires did become self-aggrandizing, they would wipe out the raison d'être for the park. The real fire revolution of the Sixties had argued to match fire with land. In many places, this meant restoring fire. In Joshua Tree it meant excluding it. The future demanded a return to the past.

Fire management plans evolved accordingly. The latest calls for aggressive suppression—catching 95 percent of all ignitions during the first burning period, this for wilderness, recreational sites, or visitor centers. Although for very different reasons, Joshua Tree is thus moving in the same direction as so much of Southern California. This kind of initial attack, however, requires an apparatus that the park cannot maintain in-house, so it has joined the larger regional consortium. Fire dispatch in fact is handled out of the Federal Interagency Communications Center on the San Bernardino National Forest. Beyond that, the park will protect structures, but by crushing vegetation, not burning it. And it will promote the old standby, "further research." Even the scientists of the Western Ecological Research Center of the U.S. Geological Survey, however, are subject to strict limitations as to the size ("minimal") and seasonality of their experimental burn plots.

Ultimately, Joshua Tree imagines a biological control over fire, not by a cycle of burning and fuels, but by the ecological stabilization of the flora. Fuel loads mean little; the exotic grasses, everything. What would seem an administrative atavism when agencies were scrambling to get fire back into the land is exactly what this particular place requires. Fire exclusion is a means to limit the invasives. Until the theory and practice of biological control matures, fire prevention remains, paradoxically, the more progressive option.

CHANNELING FIRE

The Channel Isles are a cluster of Transverse miniatures. They formed geologically from the same tectonics, they share a similar biota, and at lower sea levels during the glacial maxima, they were joined to the mainland. They have chaparral and Santa Ana winds. They have, in brief, the same traits that the fire regimes of the Transverse have. But historically they came together less intensely, and today, they come together hardly at all.[3]

Some of the distinctiveness is simply the outcome of being islands. They are surrounded by sea, which makes them more maritime; fog is more pronounced, air is more humid, and during the higher seas of the glacial minima Santa Barbara, Anacapa, and San Miguel were likely submerged outright. Unlike the Transverse, fires cannot burn into them from elsewhere. They exhibit even fewer lightning ignitions than the Santa Ynez—only three in 140 years of records. The isles are further removed from the Santa Anas. Until roughly 1840 humans set fires, although how often and in what patterns is unclear.

Then fires virtually vanished. Recently they have returned from accidents (rescue flares set 600 acres aflame on Santa Cruz) and prescribed burns, one of which escaped under Santa Ana winds. What caused the fires to disappear was the introduction of vast herds of cattle and feral flocks of sheep and goats. The abolition has been so thorough that a few plants have even shed some fire-adapted traits such as serotiny.

The isles are a borderland. Two ocean currents—two maritime regimes —merge here, much as two deserts do at Joshua Tree. But the profound border is historical. For 4,000 to perhaps 13,000 years humans have occupied the Channel Islands. The fires of record are overwhelmingly theirs; they coevolved with the biota to sculpt a distinctive fire regime, one in which large and savage fires were apparently uncommon. Then both fires and indigenes disappeared in a handful of years. Russian fur traders, hunting sea otters, probably arrived in the early 1820s. The first colonizing whites appeared on the largest island, Santa Catalina, in 1824. The indigenous Pimugnans ceased in the early 1830s, probably from a combination of disease and emigration. By the 1840s most of the larger isles had resident populations that herded cattle, sheep, and goats, along with some horses. Among the isles there were further transfers. The Wrigley family bought Santa Catalina Island in 1909 and imported exotic animals such as mule deer, black buck, bison, and boars for hunting; and by 1930 it had eliminated its sheep. Santa Rosa also got pigs, which went feral and stymied many fruiting species.

Stocking islands with domestic fauna, particularly goats, was an old practice of voyaging Europeans. The Portuguese did it with every island they discovered, letting the animals, now without predators, breed wildly and create a living larder for future mariners. Other explorers did likewise. The flocks went feral. Even deserted islands like Juan Fernandez (famous as the place where Alexander Selkirk was dumped, and the inspiration for Defoe's *Robinson Crusoe*) had goats. The long term outcome was to drastically restructure ecosystems. Islands like St. Helena became textbook exemplars for the terrors of overgrazing, deforestation, and desertification.

Slow combustion by metabolizing beasts replaced fast combustion by flame. The exotic livestock ate in such quantity that fire had little to burn, and plants that assumed a shrubby habit on the mainland now became arboreal as branches survived only above the browse line. With astonishing speed and ruthlessness the fires vanished. Here and there, from time to time, a small burn escaped, or an accidental fire started, or perhaps in the early years ranchers set fires to expand pasturage. But the old fire regime was trampled, chewed off, and eaten up. Without a change in its physical matrix—climate and terrain remained constant—the fire regime turned inside out. It was a magnificent if horrifying demonstration of biological controls. What an invasive flora did at Joshua Tree an invasive fauna did on Channel Islands.

The wreckage was self-destructive: even cattle and goats struggled to survive. By the 1950s, as the Southern California economy shifted from agriculture to services, particularly tourism, the cattle were taken out on Santa Catalina. In 1980 Congress created Channel Islands National Park, which added the charge to restore something of the natural conditions. Formal programs commenced to remove the feral fauna and reinstate a more native fire regime. But early experiments, including several under cooperative programs with the Nature Conservancy, failed. Prescribed fires burned poorly because the habitat was no longer suitable to carry flame or because the burns escaped under Santa Ana conditions. More fundamentally fire officials realized that repeated burning on the model of the southeastern United States only led to another type of conversion, from what remained of indigenous flora to annual weeds. So long had fire been excluded that portions of the biota were no longer adapted to it. On the mainland, chaparral ecology required a suitable regimen of burning. On the islands, it required an absence of flame. The only way to restore a fire regime from which fire had been extirpated was, paradoxically, to continue to exclude fire.

Fire suppression is the official policy of the islands. On Santa Catalina Island, under the jurisdiction of the Los Angeles County Fire Department, all-out control prevails. In 2007 LACFD actually sent engines to a bad fire by requisitioning a flotilla of Marine Corps hovercraft. On the other isles, where fire management is a collaborative enterprise with the Los Padres National Forest and the Nature Conservancy, the National Park Service considers all fires as wildfires and will suppress them using minimum impact suppression tactics. Both fire and fire control are subordinate to the larger park mission. Fire protection means suppressing the metabolic burning by feral beasts.

MEDIATING FIRE

Santa Monica Mountains National Recreation Area is not a distinct landscape. Joshua Tree is more or less isolated in the Mojave Desert, and the Channel Islands are outright isles. But SMMNRA is a tumbled terrain between the San Gabriels and the Santa Ynez—in the thick of the Transverse Range—and a pastiche of private and public lands assembled in 1978 without a core, thus offering a landscape mirror to the centerless city. It embraces 20 different types of landowners, some 70 stakeholder groups, and a mélange of fuels that make fires inevitable and largely unmanageable without full type conversion to exurb. If the fire scene at Joshua Tree was the result of exotic plants, and that at Channel Islands the product of exotic animals, the peculiar dynamic at Santa Monica Mountains was an outcome of exotic humans.

The mountains are crusted with chaparral over which the Santa Anas rush like water cascading through a rapids. Where gorges exist, they run roughly with the grain of the wind, creating fire flumes or the flaming equivalent of floodplains. Malibu Canyon is to wildfire what the Red River is to flooding. A Mediterranean climate, a fire-flushed biota, rude mountains, and gravity winds—if any place might demonstrate that fire is inevitable, and therefore essential, and that might illuminate the folly of arrogantly muscling suburbs into a landscape foreordained to burn explosively, the Santa Monica Mountains should be it.

When the NRA was founded, it accepted the prevailing wisdom of the day. For the Park Service this meant it should seek to restore the natural regimen of fire, or where complications from politics and development made that impossible, to use prescribed fire as a surrogate. Fuel reduction burning could quell unruly chaparral. Broadcast fire could allow a traumatized ecosystem to begin healing. These notions embodied the most progressive scientific thinking available. Research confirmed that chamise—the most emblematic of the chaparral constituency—underwent a seasonal and secular cycle of burning. The later the season, the more prone it was to burst into flame, with ideal conditions coinciding closely with Santa Ana conditions. The older it got the more likely it was to burn and to carry the rest of the complex with it. The longer fire suppression continued, the larger and more savage fires got. The solution was to turn chamise's propensity to burn against it and do the burning when people favored.

These concepts were embodied in the first fire plan, approved in 1986. By 1994 the plan received a thorough overhaul. The primary purpose of fire management, it declared, was to reinstate a more natural fire regime, but given the magnitude of "urban interface" and "private inholdings" at Santa Monica Mountains, there could be no "natural fire program," only a prescribed fire program used "in lieu of" nature's way and to "duplicate natural ecosystem processes lost as a direct result of fire suppression actions." The plan imagined a five-year program of prescribed fire to burn inherited fuelbreaks and to create "mosaics" among the brush. It refused to designate any fire management units since both suppression and prescribed fire objectives remained "constant on all NRA lands." Cynics thought the program was based in equal measure on unattainable ideals and an assumption that all actions taken would be subject to lawsuits.[4]

But the court of opinion reversed itself. Research into fire history suggested that the large, high-intensity fires for which the Santa Monicas have become notorious may be a more recent innovation. It may be that exurban development did not move into a conflagration zone so much as it created one. The offshore sedimentary record documents a long period of soot and deposition, which could only have come from fires burning under Santa Anas; but these were rare because there were few ignitions in the interior mountains. Lightning starts fires, but sparsely, and the indigenous Chumash burned often but mostly in coastal sage and grasslands, not inland chaparral; their fires would burn against the offshore winds. It was modern settlement that cracked open the mountains with roads and brought fire starters inland. The odds of such ignitions capturing a Santa Ana increased hugely. At the same time, the peculiar development of the mountains by the rich and famous made it difficult to conduct traditional burning by ranchers or to substitute other kinds of controlled fire.

The only fire allowed was wildfire. The wildfires came more frequently. Along the coast they pushed the biota into grasses, notably invasive annuals. In the mountains they shortened the cycle of chaparral burning. Fires burned out to their newly inscribed boundaries. The fire departments of Ventura County and Los Angeles County fought them. Then came the NRA. It protected the native biota, but it also, not incidentally, protected the property values of those who already resided within its capacious boundaries.

By 2000 experience and research both determined that the plan no longer worked—could never work because it was based on flawed assumptions. Prevailing wisdom now declared that to be effective mosaic burning would have to assume a scale that would consume the entire landscape; and there would never be enough money or political will to undertake such a project. Nor was it needed. The accepted view was that wind, not fuels, drove the characteristic fires and that the fire history was one of escalating, not dampening, burns. Dense stands were the norm, not an aberration created by misguided fire suppression. In fact, introducing more fire would further unhinge the system. It would spread exotic grasses, it would lessen the age of chaparral that needed to grow older, and it would surely lead to lawsuits. The best solution was to back off, try to prevent fires, and when a big one occurred, to stand aside and let the flames wash over hardened structures. The park headquarters at

Thousand Oaks, built into a hillside rather than on top of it, was probably intended to embody ecologically sensitive design features. In truth it resembles a bunker. It could survive a seven or eight on a fire Richter scale. It cast into concrete the future of fire management.

In a famous essay later included in *The Ecology of Fear* (1998) critic Mike Davis made the case "for letting Malibu burn." His argument evolved out of a sense of public economics and social justice, that it was outrageous that the entire country should subsidize through its colossal firefighting efforts the life of a financial and celebrity elite who chose to live in a fire flume. The climax was the 1993 fire siege that made Malibu burning (again) prime-time theater. The argument went unanswered. It seemed that the public was willing to pay for the spectacle, not only of panicked celebrities fretting over the safety of their horses, but of an insouciant privatizing of profits while socializing losses, and a confirmation of assumptions about life in California. Los Angeles is, as Davis noted, the place critics love to destroy. After fires again swept through in 2003 and 2007, the spectacle lost what little charm it might have had. For decades the state had relied on subsidies from the nation to support its indulgence. Now the country could no longer borrow to keep up pretenses.

Yet Santa Monica Mountains NRA did answer Davis's thesis. It said, in effect, it would no longer guarantee blanket fire control—could no longer justify it on ecological grounds, quite apart from economics or notions of social justice. It would help with hazard reduction around private inholdings; this was neighbor helping neighbor. But it would not fight fires where they were unfightable and it would not light them in the name of fire control where they would damage the greater good of the park. It made the NRA a higher good than celebrity ranches. It is an idea as radical as Mike Davis's, and it—not some hypothetical past state of nature—will likely determine the future of Santa Monica fire.

IMPERIUM IN IMPERIO

I N 2011 TWO CENTENNIALS commemorated the origins of America's modern era of fire protection. The U.S. Forest Service celebrated the Weeks Act, which created the basis for cooperative fire protection between the federal government and the states and established a national infrastructure that still allows for common practices and mutual assistance. And Los Angeles County honored the creation, after several stutter steps, of its Forestry and Fire Warden Department, which evolved into a full-spectrum fire service.[1]

When they began the two institutions had much in common. They shared a common birth parent in forestry. Gifford Pinchot, the founding chief of the USFS, was the country's first native-born forester; and he had persuaded his family to endow the School of Forestry at Yale, whose first class graduated in 1904, the year before the Transfer Act gave the Forest Service responsibility for administering the forest reserves. Stuart J. Flintham, LA County's first forester, had graduated with forestry degrees from Cornell and Yale and then worked for the USFS before moving to California. The two agencies accepted common ambitions: to protect the land from fire and to reforest what was damaged. (The California Board of Forestry completed the triumvirate.) But they also differed in ways subtle and profound.

Over time these divergences widened into a chasm. The Forest Service worried over woods and an imminent "timber famine" and saw its mission as managing for what was termed "forest influences" on the public domain. Los Angeles County fretted over brushland watershed, attempted afforestation, and saw its mission as the protection of life and property in a metastasizing metropolis. At a national level, the difference is the bifurcation in how Americans live on the land that has split the countryside into wildland and city, each with its separate fire institutions. The Forest Service had begun with wildlands and over time had to cope with an urbanizing periphery. LA County began with an urbanized core and had to absorb a fractal wildland fringe.

Accordingly, they evolved very differently, despite a common pedigree in academic forestry and a shared conundrum in urban sprawl. They viewed the scene from opposite sides of the I-zone. The USFS protects houses reluctantly since those structures reside outside its jurisdiction and its mission, which it sees as overseeing the uninhabited public estate. The Los Angeles County Fire Department protects houses because that is its primary charge, and it manages yet-uninhabited lands so far as necessary to serve that goal. The competition, and cooperation, between those two visions define much of the contemporary California fire scene.

What happened in Los Angeles County is a cameo of California fire history. Whatever their hopes and ambitions, county foresters watched their fire mission overwhelm their forestry mission, and this scenario, pioneered by LACFD, was repeated at the state and California federal levels. The metamorphosis of the California Board of Forestry into Cal-Fire recapitulates the LACFD scenario; so does the U.S. Forest Service's evolution as Region Five; and, as critics darkly worry, so might wildland fire management in the country overall. Although history is fine-grained, full of idiographic events and the quirks of personalities and places, the visible outcome can resemble a kind of remote sensing in which any heat source strong enough to be recorded saturates an entire pixel. The effect is exaggerated but can be defining. Its peculiar fire scene filled the Los Angeles County pixel; the LA County pixel filled the Southern California pixel; the Southern California pixel filled the California pixel; and many elsewhere in the country fear the California pixel may fill the American pixel.

In this way LACFD is not, in brief, simply a convenient emblem of such changes. It has been a major driver of them. It pioneered a unique hybrid of wildland and urban fire services, and thanks to the interlocking system of fire suppression it projected that invention beyond its borders.

Much of the character of California fire is an outcome of LACFD's vision, tenacity, wealth, and genius for publicity. For a century in California it has been the strange attractor, the not-always-seen but ever-present disturbance that tugs and yanks and pushes the quotidian world of fire protection.

Los Angeles County sprawls across 4,000 square miles, reaching from the South Coast beaches over the Santa Monica and the San Gabriel Mountains into the Central Valley. It embraces the major biotas, industrial ecologies, and urban landscapes of Southern California. The ontology of its fire protection system recapitulates the phylogeny of fire agencies in the state.

The county created a board of forestry that echoed the state's. Its director was titled Forester and Fire Warden. It thought in terms of forests and "forest influences," and since trees were sparse, the agency planted them, for which it quickly established a nursery. But the critical duty was fire control: fire protection for rural and small urban landscapes, for watersheds that regulated the vital water flow for town and agriculture, for the newly planted woodlands. The fire-flood cycle in the mountains was the leverage for public funding. Watershed councils contributed funds to build trails and fuelbreaks. The county distributed monies to the Angeles National Forest to do likewise. For a decade the county experienced an annual roster of 100 fires that burned 20,000 acres.[2]

It was the big fires and the bad years that mattered. The Big Ones burn off the most vegetation, set the most debris in motion, and most mobilize political will. The first crisis year was 1919 when two fires on the San Gabriels, one at 40,000 acres and the other at 75,000 acres, blew away the combined efforts of the county and federal foresters. Two years later the USFS convened the first national fire officer meeting at March Field, and two years subsequently, the state legislature allowed for the creation of fire districts, which set off a frenzy of constituted districts. Amid these rising expectations came the revolution that followed.

Everything happened in 1924. In California the fire protection districts began to gel; the country fire warden was deputized a state warden; the state contracted, with those counties that wished to participate, for fire protection on state responsibility lands. These funds became a regular budget item for the county. Essentially overnight, Los Angeles County acquired the third-largest fire department in the state. Meanwhile, Congress passed the Clarke-McNary Act, which greatly expanded the scope of the 1911 Weeks Act to allow for forested watersheds that might not contribute to navigable rivers; a trailing amendment specifically included the "brushland" watersheds of Southern California. The spigot opened for federal funds to the state. Since the state paid Los Angeles County to provide protection, that money flowed to county coffers. In fact, LA County claimed the lion's share of state monies. At the same time, the county and the Angeles National Forest signed a mutual aid agreement that allowed each agency to make initial attack as needed across their mutual border and for reimbursement after the fire was out. The arrangement sparked one of the great tag teams in fire protection as the two agencies found ways to complement one another and compete for innovations. Within two weeks a wave of fires swept over California—the first fire siege, really, of the modern era. Over 50,000 acres burned on the San Gabriels. The Forest Service was so alarmed that it convened a national board to review the season, the initial installment in what would become a California tradition.

The outcome confirmed the basic institutional infrastructure for fire protection. Within weeks the largest county fire department in America—one of the biggest of any jurisdiction—sprang into existence. That it appeared to come from nowhere and crystallized, under crisis, almost instantly seemed peculiarly and suitably Californian.

As it emerged from this forced chrysalis, fire protection in Los Angeles County found itself split into two distinct operations. One was anchored in the fire districts, which matured into urban fire models as their landscapes ripened into cities. They were funded by property taxes; they used regular engines and painted them red; fire crews dressed in black uniforms. The other operation was the forestry or mountain division. This was a wildland fire brigade, more or less interchangeable with U.S. Forest Service or California Division of Forestry crews. It relied on county appropriations supplemented by state and federal funds. It ran foot crews and pack strings, and where vehicles were possible it invented hybrid pumpers, which it painted green. Crews dressed in light green work clothes. There was no way to confuse one group with the other. The county, in fact, exactly reflected the state overall. All that joined them was the informal sobriquet, the Los Angeles County Fire Department. There was no other easy way to fuse the two systems.

As the decades passed, the red overtook the green. The land was paved and pocked with houses; the nominal wildlands were crowded to the margins or left over the hills; industrial accidents challenged wildfires for the headlines. Highway accidents rose. Lookout towers came down. The money, the political interest, the future—all slowly turned red. Crises helped. When the San Francisquito Dam broke, when an earthquake shook Long Beach, when a natural gas pump erupted into a fountain of flame at the Santa Fe Spring oil fields, when the Pickens fire was followed by the LaCrescenta-Montrose flood, the county turned to its fire department as the only agency capable of tackling the required logistics. During the Depression, when labor became cheaper, the forestry division continued to innovate by using the CCC and then creating comparable camps for juvenile inmates who were trained to fight fire, a program subsequently expanded to include adult inmates. In 1940, as war clouds blackened, the fire chiefs of the Greater Los Angeles area gathered together to find ways to share resources during emergencies. The upshot was a Mutual Aid Act, which was subsequently expanded throughout California by the state legislature in 1941. During World War II, as labor dried up and war-industry growth overflowed, LACFD turned, as it always had, to technology and cooperative programs to pick up the slack. It made the case that it was appropriate to externalize funding—to insist on outside contributions—since the war was a national undertaking.

The postwar boom brought further reorganizations as urban sprawl outpaced fire protection, as it did most other services. Each consolidation brought the forestry division into greater harmony in structure and style with the general fire department. By 1954 the two divisions became, for administrative purposes, interchangeable. That year bad fires and floods led to an interagency Southern California Watershed Council that sought to balance funding for the mountains. Usefully, the burns coincided with Operation Firestop, a multiagency project aimed at transferring wartime science and technology to firefighting (LACFD was a major sponsor). By now LACFD had responsibility for half of a county urbanizing at breakneck speed. (In July Disneyland officially opened amid former orange groves.) The next two months saw breakout fires that exceeded those of 1954 and began killing crews. In 1956 it was Malibu's turn (again). In 1959 the Woodwardia fire, started by a disgruntled firefighter, so alarmed county officials that the board of supervisors declared LACFD Chief Keith Klinger a "fire czar," free to order whatever he thought necessary. The fire claimed the first death by an air tanker drop and the first air tanker fatality. The two realms of fire, open flame and internal combustion, were converging with thoughtless, lethal power.

But much as the landscape was rapidly urbanizing, so was the fire service. Between 1954 and 1969 personnel had doubled to 2,500, and stations had multiplied from 80 to 113. All divisions shared this explosive growth. So fast had the region sprawled, however, that LACFD had to rely on technology rather than staffing to meet the demand, particularly for wildland fire for which engines could not substitute, as they could in urban settings, for handcrews. The other strategy, also traditional, was to find cooperators who could fill needs during emergencies. The state's master mutual aid agreement, overseen by the Office of Emergency Services, was one solution; and after the 1970 season, so was Firescope. LACFD was among the Big Five fire agencies who signed on to the Firescope prototype. When the program went operational in 1976 to 1977, it was headquartered in LA County.

Behind the move to consolidate—an old theme for LACFD—was a worsening fiscal crisis. Even as development blew explosively outward, funding was imploding. Legislation in 1972, 1976, and especially 1978 with Proposition 13 gutted the property tax that financed fire districts and contributed to the county general fund. The federal Cooperative Fire Program also underwent a contraction that in principle (though not always in practice) deleted the generous grants to state forestry bureaus. In response the county consolidated further, hugely in 1990, and found ways to draft some lost monies from the state treasury. What had emerged as a clever practice—mutual aid—became a necessary one.

The big fires kept coming. Flames broke out; crews fought them and sometimes fell. Malibu Canyon was a fire bellows. Fire sieges gripped the South Coast in 1993, 2003, and 2007. The first two came on the heels of economic busts that followed the end of the Cold War and the dot-com bubble; the last coincided with the subprime real estate bubble that burst to become the Great Recession. Yet while property values may have tanked, they were still substantial: LA County embraced a wealthy part of the largest metropolis in a wealthy country. The assets under protection were immense; and the future promised more development, which would demand yet more intensive shielding. LACFD rose and fell with the regional economy generally. Or it did until the state and federal governments found they could no longer afford to subsidize California in the manner to which it had become accustomed. Only the war on fire, a conflict without a horizon, endured.

When the smoke lifted, LACFD was an urban fire service—an all-risk first-response outfit—with some peculiar landscaping in its jurisdiction. Fewer than 20 percent of its alarms involve fires of any kind. The forestry division shrank to a relict appendage. Although some 42 to 43 "badged foresters" remain on staff, their number is dwarfed by the scale of the agency overall, and they serve an urban constituency. Today, LACFD supplies fire services for 58 of the county's 88 cities. It has responsibility for the county's unincorporated areas, which it protects under contract from CalFire. It is the regional coordinator for California's Emergency Management Agency. It has mutual aid agreements with Santa Monica Mountains National Recreation Area and the Angeles National Forest. It is one of two departments qualified for international deployment for urban search and rescue. It is one of the five largest fire departments in the nation.

Its story, however, is more than big money protecting big-money assets. On the national scene its landscape and mission made LACFD an outlier, but an outlier that paradoxically claimed a nuclear core of American fire. The Los Angeles scene was not simply an index of national trends but an instigator of them.

Both LACFD and the Angeles National Forest pride themselves as agencies of "firsts." The Angeles was the first forest established for urban watershed, the first to house an experimental forest devoted to brushland, the first to devise preattack plans, among the first to create hotshot crews, exploit helicopters, helitanks, and helijumping, the first to implement the incident command system. It hosted the first national-level postseason review. It pioneered fuelbreak systems. It was an urban-edge forest before the term was invented. But LACFD has been no less innovative; it is as proactive in fire suppression as Florida counterparts are in prescribed fire. From its origins it has turned to technology as the only means to cover expansive lands with explosive fires. The department used horse patrols and mule strings, invented a mountain fire truck, sent out fire guards on motorcycles, early turned to field telephones and radio, experimented with portable pumps and flamethrowers, adapted helicopters for night flying, flew pony blimps for reconnaissance, and tested air tankers of all sizes. It was an active sponsor of Operation Firestop. It fielded bulldozers big enough to swallow Abrams tanks and outfitted them with protective cabs. When fire broke out on Santa Catalina Island, it requisitioned hovercraft from the Marine Corps base at Camp Pendleton and sent them across the strait bulging with engines. It was a first mover for inmate crews. It field-tested ideas about defensible space in the form of brush clearance regulations. It flung a net of remote automated weather stations across the county. It had more firepower than many state forestry bureaus.

It was no less innovative with institutional technics. The latent disasters that lurked in its elemental matrix of air, water, earth, and fire meant that LACFD had to be self-contained. It needed the capacity to rebuild its own capabilities if it was to assist in a catastrophe, so it has its own corps of mechanics, plumbers, electricians, carpenters, and emergency generators; it can repair bulldozers, engines, and a fleet of helicopters. But its real reach extends much further because of its willingness to seek out mutual aid and use it for leverage. Its ancient alliance with the Angeles boosted both agencies; together they dominate the region. Its contract relationship to CalFire made it a state power. That triumvirate transformed Los Angeles County into a national presence. Paradoxically, its wildlands kept it a *fire* department, even as its urban disaster capabilities have sent crews to earthquake-blasted sites in Haiti, New Zealand, and Japan.

Its story is, by and large, the modern California story in miniature. Its de novo origins, begun rapidly from scratch; its "instant" creation as a fire department by a fiat of reorganization; its blistering postwar growth; its appeal to technological solutions amid a polyglot population that both swelled and turned over seemingly as often as El Niños returned; in short, an institution that constantly reinvented itself even as it metastasized—all this looks a lot like California overall. But then one reason why California fire looks the way it does is the irrefutable presence of Los Angeles County.

In 1995 LACFD had a budget of $500 million. After two booms and two busts, it operated with a "strapped" but still daunting 2011 annual budget of $900 million. Its Forest Service sibling, the Angeles, has the largest fire budget in the national forest system. Its state partner, CalFire, has the largest state fire program in the country. Together they define an institutional matrix for the Southern California fire scene and one of the three prevailing national cultures of wildland fire. What makes the complex compelling, however, is that Los Angeles County has not been merely the beneficiary of state and federal attention (and largesse) but also a stimulant to them. Los Angeles County is the fire that won't die.

MENDING FIREWALLS

I N SAN DIEGO COUNTY the landscape softens. The mountains are less towering, the valleys less yawning, the contrast between urban and wild less shrill. Ownership is dappled; ranches and farms still spread over the higher plateau. Oak savannas and soft chaparral mingle with hard brush and granite boulders that appear to graze on the land like sheep. Orchards and parks daub the peri-urban scene. But as the urban pockets gel and add layers like a concreting nodule, the borders become more rigid, and a creeping suburban pastoralism hardens into a fractal frontier with little to buffer the once rural from the fast urbanizing.[1]

This kind of geography frustrates management, if you think of management as the application of explicit principles to distinctly bounded lands. Or as Robert Frost famously put it, good fences make good neighbors. As landscape geometry, however, San Diego belongs more with Schrödinger's cat than with Euclid's *Elements*; administratively, it requires a conceptual leap, like moving from Newton's laws to quantum mechanics. The world tends to seep between those borders, to replace the exactitude of a drawn digital world with the sloppy analogue of the real one. Moreover, borderlands are notorious as places in themselves, not simply a swirl of what fronts them. As Frost also noted, something there is that doesn't love a wall. That wisdom—both sides of it—applies to fire. And it manifests itself, with paroxysms of violence, in Southern California.[2]

Certainly fire both respects and ignores walls. The essence of fire suppression is to erect one, called a fireline; and the goal of presuppression, to create a fireline in advance, called a fuelbreak. Yet vigorous fires frequently leap over them. Some walls are both necessary and, when most needed, frequently useless. In this regard they reflect the most fundamental of conundrums in fire management, how to cope with the big fire, the rare but catastrophic event. But they also exhibit, in miniature, the dilemma of fire in the Southern California I-zone, in which society tries to mix what doesn't want to be together and to segregate what wants to join. That tension focuses on the line in the dirt intended to separate what, increasingly, has no separation. The fuelbreak, or its most recent avatar, defensible space, is where the battle has joined.

———————

The fuelbreak has a long history in the region. Local authorities were advocating (and financing) fuelbreaks as dual-purpose firelines and trails as soon as the mountains were gazetted as forest reserves. There were fuelbreaks in the San Gabriels before the Forest Service took over their administration. They provided, in principle, a means of access, a method to break up continuous fuels, a rudimentary fireline ready to activate when needed, and a visible display of administrative resolve. They inscribed their message across the great screen of the mountains. Unsurprisingly, the nation's grandest experiment was of course California's Ponderosa Way.

Over the years fuelbreaks have displayed a cycle of senescence and regeneration much like that of their surrounding vegetation. After each disastrous fire season, existing fuelbreaks are scraped clean and widened and the system expanded. Then, they decay. They are expensive to maintain; other needs clamor for the money; critics scorn the ridgeline scratchings, which they regard as ugly and useless. The secondary system overgrows. Only the primary roads and those deemed most essential receive maintenance. Then the flames rush over the landscape, the public demands protection, and the fuelbreaks return. The life cycle of fuelbreaks, in brief, shows the same rhythm as the chaparral in which they are embedded.

In classic theory the fuelbreak assumes two forms. One breaks up the interior of the reserve in order to assist fire control. The principle is identical to that used in the built environment: create firewalls that retard spread and give firefighters time to wrestle the blaze into control. If sited in a forest plantation, the fuelbreaks would be incorporated into the design of planting. The most successful outside planted landscapes act as levees rather than dams; they flank the terrain-and-wind-directed flow of the flames rather than try to stand against them. The other variety of fuelbreak is intended to guard a reserve's perimeter; it serves as a biotic or fiery fence. This is rarely an integral feature, but rather something imposed onto and across a landscape that wants to behave differently. The reserve's borders might cross firesheds as they do watersheds, which means fighting the order of nature rather than working with it.

While the arrangement of the fuelbreak system has remained more or less permanent, its purpose has changed. The reason for persistence is easy to explain: this is where fire behavior and fire control—and politics—argue for breaks, and such considerations don't alter. But the nature of the fire threat does change. Originally, foresters recognized that the surest way to prevent big (and costly) fires was to contain small ones, and for this a network of access and buffers was necessary. Originally, too, they constructed a system to prevent the promiscuous (and sometimes malicious) fires set all around the reserve from entering.

But as the belief took hold that some fires were inevitable and good and that fire control could itself be damaging, the pressure for those interior fuelbreaks lessened. They became more an act of political posturing, a signal to the public that fire officers were wary and prepared. The bigger change involved the perimeter fuelbreaks. Increasingly, these became less critical for keeping fire out than for keeping it in. The catastrophic fires were those that kindled in the mountain reserves and then spilled out onto the surrounding countryside or, more properly, cityscape. The fuelbreak evolved away from a matter of protecting the reserve from the community and toward protecting the community from the reserve. It became a question of public safety, like debris dams and flood control channels.

It was one matter, however, to sculpt a fuelbreak along a continuous border across the flank of a mountain. It was quite another when the "border" was a speckled landscape of inholdings, fractal suburbs, and infrastructure nodes, or when, amid developed landscapes, it assumed quasi-natural features in the form of parks, greenbelts, and zoning for protected species that did the speckling. Critics bridled that by the time each enclave had its belt of clearing, there would be nothing left. The issue is particularly acute for San Diego because the city has 900 linear miles of ravines and some 55,000 houses along a fire-exposed fringe. If each entity put in the 100-foot or more clearances recommended by fire authorities, the practice would effectively wipe out relict landscapes. It would convert nature to city by stealth. It would, so the argument goes, create an urban desert visible from space.

Instead, critics want to focus on the structure itself, the home ignition zone, which is to say, the structure and its immediate environs. This is where surface fires and ember attacks actually take out houses, and it is often the sheer congestion of houses that carries the fire as each involved structure ignites those around it. The critics want fire protection to concentrate its hardening here, which would shrink the penumbral zone of clearing. If development is dispersed, flames would wash around the structures, and if concentrated, they would dissipate as they ran out of fuel. A wildland fire, or an urban conflagration, would shatter into manageable house fires and then dissolve.

Behind this conception are related disputes about what fire management means. It begins with an argument about the nature of protection, or to state the issue slightly sideways, do you manage fire at its source or its sink? The fire-as-source group sees the problem as managing the landscape where fire originates. The fire-as-sink group targets the places where fire strikes. The hydrologic equivalent is whether to control flooding by improving watersheds or by erecting levees and dams. The fire-as-sink group envisions the task as focusing on the assets at risk, both hardening individual structures and making the I-zone overall more resilient. It reasons that if the goal is to protect houses and the citizens who reside within them, then protect those houses directly, not remake the amorphous landscape from which the fires emanate in the vague hope of eliminating risk. In this view nature needs protection from overzealous and misdirected fire control. They each define mitigation differently. The source group focuses on the need to mitigate against fire; the sink group, to mitigate against unwonted fire prevention measures.

Each sees with different eyes the border between the fire and house, and accordingly each assesses fuelbreaks differently. The source group, intent on managing fire within its fireshed—not only to prevent escapes but for ecological purposes—wants interior fuelbreaks to help and perimeter fuelbreaks to halt as much as possible. It can cite an honor roll of successes. The Harris fire, where burning out along the International

Fuelbreak spared Tecate, Mexico. The Border 16 fire, where flames crossed north over the border and took only a solitary house, the sole structure without defensible space. The Shockey fire, where a treated neighborhood in Campo shook off even Santa Ana–driven flames. The Banner and Angel fires, where burnouts along the Sunrise Fuelbreak shielded the town of Julian.

The sink group poses a counternarrative. It sees fuelbreaks as ineffective when they are most required. The region's vaunted defenses did nothing against the monster fires such as the Cedar and Witch that define the contemporary scene. Worse, they aggravate the abuses lavished on an already damaged landscape by creating grassy fuses to carry fire and channel invasive pyrophytes into unscathed chaparral. Giving every homeowner a personal fuelbreak in the form of overly expansive clearances called defensible space guts any hope of preserving a vestige of native plants and habitat. It means subversive urbanization by other means.

So the debate continues, each side more effective at criticizing the other than promoting their own agenda. Both groups, however, design with a particular fire and environmental risk in mind. But the Southern California scene is nothing if not dynamic. The representative fires keep morphing. The assets at risk keep moving. The only constant is the argument of neighbors across the wall, one rebuilding the wall and the other willing to let it decay.

Then the next conflagration comes.

━━━

Each side can point to failures and successes. Yet in the San Diego region there are two examples of enduring fuelbreaks that seem to perform exactly as designed. It's worth pausing to examine how and why they work and what their costs might be.

Fuelbreaks succeed best when they are integral to a built landscape, or when they are part of a planted agricultural complex. They have worked in pine plantations on the shores of the Baltic, the sand dunes of the Landes, and the Sand Hills of Nebraska. They have contained fires along railway lines, often combined with grazing, planting, and controlled burning. As greenbelts, they have shielded new communities, even cities. They have worked in India and Ghana to define and defend the boundaries of gazetted forests. In most instances the threatening fires do not scale up to conflagrations, and in no cases do fuelbreaks succeed by themselves, any more than firewalls will keep a building from burning down; but they buy time and assist firefighting.

They struggle when retrofitted or imposed over landscapes in defiance of terrain, wind, and fuels. When local conditions favor large fires, only very large fuelbreaks can help check them, and that effectively means type conversion, transforming whole landscapes, which in Southern California means housing tracts. Still, under less than extreme conditions, they can leverage fire suppression, helping channel a fire; and in miniature forms, as defensible space, they can encourage engine crews to stay with a house that they might otherwise yield to the flames. If fuelbreaks fail during extreme conditions, so do all the other strategies and maneuvers of fire management. The Big One continues to haunt wildland fire management.

There are, however, two exceptions, although they may prove the rule because they show what a strategy of fuelbreaks can cost. One is Camp Pendleton. On a map of regional fuelbreaks Pendleton's dense network sits like the textured surface of a grenade. The camp is laced with roads and fuelbreaks, all mandatory to contain the burns kindled by endless live-fire exercises. But the camp is also surrounded by a defensive belt, like a demilitarized zone, that is annually cleared and burned. For what it is designed to do, the fuelbreak system works. It is integral, dense, comprehensive. Very few fires leave Pendleton, very few enter.

The other example is the International Fuelbreak that spans some 40 miles of the border with Mexico. It originated, as so many large fuelbreak projects did, in the 1930s when the CCC program demanded public works commensurate with vast pools of unskilled labor. When the CCC left, so did the fuelbreak. It was revived in the 1950s by the California Division of Forestry, this time using California Department of Corrections conservation camp labor. It decayed again. It revived forcibly during the 1990s when Operation Gatekeeper sought to control illegal immigration from Mexico around Tijuana. The flow of migrants moved east. Abandoned campfires pocked the chaparral, escaped fires swarmed over the mountains.

The International Fuelbreak then joined other efforts to secure America's southern frontier. In 2002 an interagency alliance, using conservation camp labor, reconstituted the project. It began with a stretch two miles long and 300 feet wide, "with some vegetative islands for wildlife and aesthetics." Plans called for extending the reach through the mountains for 30 miles. In the 2003 fire siege, with the Otay fire driving south and west under Santa Anas, the initial attack incident commander successfully burned out along the break and controlled the larger burn "with very limited resources."[3]

Limited on the fireline, but not limited socially. Such fuelbreaks are a significant cultural investment; they have to tap national funds and purposes; fire alone is an insufficient justification. A conflagration could rekindle their clearing but only some larger ambition could sustain them. And beyond apathy, or a distracted public, they carry environmental costs. Even with bottomless CCC labor on hand, the Forest Service experimented with chemicals to keep the breaks clear. It had enrollees spread arsenic, and later, when CDF oversaw the network, agencies sprayed an herbicide that achieved notoriety in Vietnam as Agent Orange. The breaks became corridors for weeds and invasives.

Even a dense network still requires an active firefighting force; engines and fuelbreaks have to act in concert since protection is a social dynamic, not an inert piece of infrastructure. To commit to a model of fire control by fuelbreaks and engines, the threat from fire has to be high, for such landscapes can resemble the fire equivalent of a police state.

In 1989, four years before the cycle of Southern California conflagrations renewed, John McPhee published an essay that captured exactly the absurdist outrage that most of the country felt toward the region. "Los Angeles Against the Mountains" described the Sisyphean task of holding back the debris the San Gabriels continually shed, like a snake sloughing old skin. As with most Southern California stories this one involves fire.

The ecological narrative is simple. A big fire is followed by heavy rain that sends unstable regolith downstream. In extreme events the runoff gushes as a debris flow, as though a ridge liquefied and roiled down canyon full of mud and boulders. The social narrative is equally simple and predictable. It begins with debris dams constructed along the alluvial fans to stop the nuisance flows. The major debris events overrun them and fill them, which means the old ones must be cleaned out and new ones built farther upstream. The process escalates. The defining consideration, however, is what to do with the debris. With the landscape downstream built out—the alluvial fan has become a cone of tract homes and mansions—there is nowhere to dump the fresh debris, which means the dams must fail. The solution is an "elegant absurdity" by which the San Gabriel Dam, first erected in the 1930s, must be continually cleaned out and the only place to haul the fill is farther up canyon. "They take the debris and carry it back into the mountains," as a spokesman for the Department of Public Works explained to McPhee, "where they create a potential debris flow."[4]

Yet this is much the same formula for fire. Substitute fuelbreaks for debris dams, fuel for debris, conflagration for debris flow, all for the protection of private houses so they can survive in harm's way only because the cost of protection is absorbed by the public. Rather than fix behavior, in this case the real estate market, the region invests public monies (and national insurance funds), much of it from outside California, into infrastructure. The debris—fuel—regrows, and either it is mechanically cleared or it burns.

Its fuelbreaks thus do for Southern California what the state master plan for water does. It allows a city sited for economic reasons to hold out against prevailing natural forces. When its economy or its social resolve falters, the sea, for the one, and fire, for the other, rush in. Over and over, the problem repeats with a slight increment of intensity added to each cycle.

So it all continues. The fuelbreaks reappear in the same places. They cause the same disturbances. They boost firefighting capabilities against the lesser burns. They fail during the conflagrations. They reappear under new names or are modified to accommodate new realities—housing tracts rather than ranches, McMansions rather than barns. While in San Diego the contrast between mountain and plain is less intense, its discourse is just as fierce.

Apologists explain that anything built can only meet reasonable standards, not everything imaginable. Engineers design for a 50- or 100-year flood, not a millennial one, or for a 5.8 or 6.7 earthquake, not for a

Richter 8. Similarly, fire agencies traditionally plan for an average worst event. But Southern California operates on extremes, not means; there is no "average worst." When it breaks, it tends to break completely. You might be able to stop the flow of flames but not the shower of sparks. For every reason advanced to keep fuelbreaks, there is a reasonable objection, and a reasonable objection to the objection, like debris dams backing up debris dams. The pragmatic solution would be to shift the argument about whether, abstractly, fuelbreaks are right or wrong, to whether, practically, in particular places, they are useful or not.

The real reason the discourse endures is because those wildland firewalls are tangible emblems of philosophical differences. They trace out separate conceptions of fire and its purpose, they define the border between competing ideas about how to live on the land. So even as they are scraped down and grow back, the old lines of debate persist, like buried fault lines that from time to time rupture under the stresses of deeper forces. Those walls are, in the end, the lines of negotiation between a nature that doesn't care and a society that doesn't want to worry.

FORCE MAJEURE

T RANQUILLON RIDGE juts bluntly out into the Pacific to make Point Arguello, as the last projection of the Santa Ynez Mountains, themselves the westward outlier of the Transverse Range. There it acts as a geographic hinge. Below, the scalloped California coast runs mostly east and west; above, it veers northward. Sprawled over Tranquillon, Vandenberg Air Force Base (VAFB) lies at the pivot. It claims just under 100,000 acres that embrace the coast south of the ridge along with the grassed and hilly valley of the Santa Ynez River to the north. It is an ideal location for VAFB's primary mission, the launching and tracking of polar orbit satellites. Any launch failure will scatter its errant fire over the Pacific.[1]

It would seem, as well, an ideal place to test theories of Southern California fire. Half the base has no landed neighbor. The most powerful winds blow offshore. The only urban complex lies within the base on manageable floodplains. While subject to the same constraints as federal civilian agencies, VAFB can appeal to a higher cause that might overcome the various obstacles to wildland fire management. Vandenberg has tested Minuteman missiles and the nation's first nuclear-tipped ICBMs, and it launches most of the surveillance satellites critical to the national intelligence mission. Surely, national security, backed by military might, not least budgetary, would be a force majeure that could blast through the impediments that seem to litter the California landscape with institutional caltrops.

The reality is that even Vandenberg hinges on the informing traits of Southern California fire. Instead of liberating fire management from bureaucratic nettles and vines, its military mission has added to the thicket. Wildland fire management at America's Gateway to Space looks much like fire protection in the frontcountry of the Los Padres National Forest and Santa Barbara County. Even the U.S. Air Force Space Command has had to yield and has hammered its swords into brush hooks.

The western Santa Ynez Mountains, ending in Tranquillon Ridge, have a long history of human habitation and fire. The landscape overflows with relics of Chumash occupation, and no doubt their fires littered the landscape as fully. The missions broke that pattern, eventually replacing them with a mosaic of valley cultivation and mountain grazing.

For the next 200 years the Point Arguello region underwent a series of economic type conversions that perpetuated the pattern first established by Spanish friars who organized the land into big-estate holdings, an arrangement that left large open spaces vulnerable to large fires. Ranchers continued to burn until the War Department acquired 86,000 acres from them in 1941 to establish Camp Cooke. Further purchases boosted that mass to 94,000. The camp trained artillery units and eventually housed German and Italian POWs. When the war ended, the land regrew to its old habitat, reverting to grazing leases and patchy farms. Then the Korean War broke out, and the camp was reopened for combat training; it shuttered again in 1953. A few months before the Soviets launched Sputnik, Camp Cooke was again reclaimed. Its southern extent became the Naval Missile Facility at Point Arguello. Its northern portions went to the Air Force's Strategic Air Command to become Vandenberg Air Force Base, while its disciplinary barracks were transferred to the Bureau of Prisons. In 1963 an administrative restructuring assigned the navy base to Vandenberg, while land acquisitions, ending in condemnation proceedings, added ranchlands to the south, principally the Sudden Ranch, to provide clear flight paths for polar-orbit launches. By 1969 Vandenberg had reached its final dimensions, 99,099 acres.

The pattern of land use had eerily resembled a series of large burns. Each reclamation cleared a significant chunk of the landscape, which then grew back to something like its former state. Syncopated with those constructed clearings, ranchers, accidents, and ordnance continued to fire the land. The last major outbreaks occurred during its era as a live-ammunition training grounds during the 1940s and early 1950s. After that, most fires were small and unrecorded; the bigger ones had to dance around the dappled old fields, manzanita, bishop pine, and wetlands. Much of the year the prevailing northwest winds bathed the base with fog (during an acute episode in 1923 seven navy destroyers had run aground at Point Arguello). For a while, too, the old ignitions faded along with the ranchers. What replaced them were faulty power lines, errant missiles, and other ephemera of an industrial society. The base erected a modern fire service to control fires among its facilities and the vacant lot lands between them.

For nearly 20 years after Vandenberg had been commissioned as the Western Space and Missile Center, there were few wildland fires, only one of any consequence, which burned on the northerly slopes of Tranquillon Ridge. About 15 percent of the base's landed estate was actually developed. Most was quasi-native flora. Security on base was tight. The fuels that mattered were the solid propellants stored at space launch complexes (SLCs). The fires of eminent concern were those that ignited Minutemen, Atlases, and Titans.

In 1977 California had endured almost three years of intense drought, and that summer brought a mob of fires that, in the fall, joined up with Santa Anas and rioted from Oregon to the Mexican border. By December fire authorities believed the siege had ended. Instead, on December 20, it stormed Vandenberg.[2]

South Base includes Tranquillon Ridge, and apart from a few SLCs, it is both open and hilly. Running east to west with the grain of the ridge, Honda Canyon plunges from the summit to the seacoast. Roads that serve as fuelbreaks flank the two confining ridgelines. Honda Canyon resembles a rough-hewn sluice box, and when the offshore winds blow, they rush down it like a flume. The early winds blew down a power line that crossed the gorge, a spark took, and the first-responding engine crew stared down the steep slope. The winds strengthened and along the crest, at times, exceeded 100 miles an hour. The Santa Ana had become effectively a dry category two hurricane. It whipped the originating flame into fiery whitecaps and then a storm surge that poured down the canyon. The Vandenberg fire service mustered, called in mutual aid, and attacked the fire along the ridgelines. Their forces were overrun.

Before the day ended some eight crews suffered burnovers. The fire flashed over Engine 12 at 8:56 a.m.; a mobile fire command post at 9:36 a.m.; a Santa Barbara County strike team, VAFB's Engine 11, and a security police vehicle at 9:40 a.m.; Ambulance 3, then transporting a burned dozer operator, at 10:10 a.m.; a VAFB and Santa Barbara County dozer team at 10:55 a.m.; the Titan SLC-04 complex at 11 a.m.; a VAFB dozer and rescue truck at 2 p.m.; and two VAFB firefighters on the Coast Road at 2:30 p.m. The second burnover not only proved fatal, it obliterated the fire command structure by killing the base commander, Colonel Joseph Turner, Fire Chief Billy Bell, and Assistant Fire Chief Eugene Cooper, while simultaneously burning over the vehicle containing Santa Barbara Battalion Chief Don Perry and a hotshot crew superintendent, Joe Lindaman. The Santa Barbara group hunkered down in their sedan, aided by a fire shelter. The Vandenberg command bolted from their Chevy Suburban and died in the scrub. The fire dissolved into a pandemonium of flame, smoke, and engines and dozers scattered like embers before a howling wind.

The military instinct was to reestablish command, communications, and control, then regroup, rally, and counterattack. General David Gray knew a crisis when he saw one and did what training and experience required. But while reasserting a clear chain of command, calling in reserves and marines, ordering more dozer armor from navy Seabees, and demanding that crews dig in around critical assets might work against a sudden attack, you don't "fight" flames under Santa Anas. You don't throw "augmentees" into the line of fire. Digging in means little when winds can hurl firebrands miles beyond your position. A firefight means letting the flames run and holding the flanks with burnouts—a tactic that to military minds can seem like lobbing artillery into your own positions.

At the Tranquillon hinge, three ecologies of command converged, one military and two fire. The military ruled—it was the highest authority, and it understood how to fight a war and oversee a crisis, though not how to directly confront a wildfire. The VAFB fire organization was particularly attuned to the built environment, tweaked for special hazards from launch complexes and from chaparral-encrusted mountains. The launch landscape held horrors like SLC-6 stockpiled with liquid rocket propellants and solid fuels that, if ignited, could blow a crater five miles wide. The wildland fire organization came mostly from outside through the medium of the Santa Barbara County Fire Department. They spoke different languages, but even translation was troublesome because they literally could not talk over a common radio system. VAFB had its own emergency radio channel, the wildland fire units streaming into South Vandenberg had another; neither could speak directly to the other. Eventually the brass yielded tactical decisions to its fire officers, the fire blew out to the sea and spread out like a debris fan. Then the rains came, breaking the fire and the drought.

The Honda Canyon fire killed four men (a dozer operator, Clarence McCauley, subsequently died from burns). It might have taken dozens, and although the launch facilities, and fuel storage tanks, survived unscathed, the flames might have fatally compromised Vandenberg's larger mission. Out of the ashes emerged VAFB's modern fire management program. Perhaps surprisingly, perhaps not, it looks like nothing so much as the standard Southern California operation modified for a peculiar built landscape. Even the Pentagon had to bend to the regional character of fire, even when that regional outlier jutted into the Pacific.

Vandenberg commissioned studies, reorganized its on-base fire operations, and moved to join the Southern California fire community as a full participant. Without integration, Honda Canyon would surely be the first in a wave train of lethal conflagrations. But the more VAFB integrated, the more it had to submit to the logic, rules, and style of fire management of its time and place. It came to resemble every other member in its fundamentals.

One reform targeted better understanding of what, exactly, the fire threat was. South Vandenberg had been acquired, by legal force, to assist launches, without regard to its environmental character other than as open space. But "open" didn't mean "empty"; and vacant land, once freed from routine grazing, burning, howitzer blasting, and maneuvers grew

lustily. By the mid-1970s most of the base had at least 20 years of revan-
chist chaparral, and much of South Vandenberg another decade or two.
That brush was not inert ground, like a posted target. It could burn, and
the longer the span of time, the more likely that fuel would find a spark
and wind to power a conflagration.

After the Honda Canyon disaster, the base entered into a cooperative agreement with the U.S. Forest Service and the Department of Forestry and Range Management of the University of California, Berkeley.
Their findings were published in 1980 as an environmental analysis report
that included an operational "wildland fuel management" plan. The gist
reflected the progressive thinking of the times, which sought to shift
fire protection from an emphasis on firefighting to fuel management.
Through a network of strategic fuelbreaks and prescribed burns, VAFB
could contain the wildfire hazard while satisfying its obligations to protect ecological assets as well as launch complexes. A 1994 agreement with
the University of California, Santa Barbara, accented those biotic benefits, and the need to reconcile fire protection with legal requirements to
manage for archaeological resources, threatened and endangered species,
air quality, and other habitat considerations.

Meanwhile, the base needed more boots on the ground. It created a
specialty fire crew led by military staff, a wildland version of what it had
for structure protection. In 1980 the crew morphed into a full-blown hotshot crew (still, the Department of Defense had to write the job specs
according to structural firefighting). The Vandenberg Hotshots made the
base into a regional presence with an Interagency Training Center, and
eventually, a national one for DOD. It assumed training responsibilities
for all defense wildland fire support, much as Florida's Eglin AFB pioneered prescribed fire. That Vandenberg is the only DOD facility with a
legitimate hotshot crew speaks volumes about the power of the Southern California fire scene to mold institutions, or more aptly, about how
a particular fire culture can propagate. Meanwhile, the Department of
Defense steadily integrated with the national fire community by sitting
on the National Wildfire Coordinating Group, posting a liaison at the
National Interagency Fire Center, and accepting national fire qualifica-
tion standards. By affiliating with the Southern California community,
Vandenberg was already ahead of that front.

The pressures extended to prescribed fire. The original scheme proposed controlled burning to dampen the volatile chaparral—this reflected the regional consensus, based on research into chamise, that fuel age classes were the critical factor in structuring fire regimes. For a while the burns proceeded; they were useful for training as well as for fire protection the way urban fire services might burn down vacant houses. Gradually, however, broadcast fires met resistance until they shriveled into highly contained spot burns around launch facilities or in patches crushed or masticated, the equivalent to burning windrows. This was fuel processing, not ecological burning.

In fact, despite the antiquity of fire, it increasingly came to be seen as an environmental threat as much as a risk to the built landscape. The Vandenberg environmental group, which must approve all burns, continually finds new reasons to limit the fires. The base has 13 threatened and endangered species, it has remnant patches of tan oak and bishop pine, its coastal sage is judged as overburned, it has splashes of wetlands that host fairy shrimp and California red-legged frogs, and even old-growth chaparral draped with moss found nowhere else. There are invasive exotics ready to seize disturbed sites, from planted eucalypts to South African ice plant. There are middens of artifacts from Chumas times. Large swathes of South Base are laden with unexploded ordnance from anti-tank land mines to grenades to mortar shells, all invisible amid the dense shrubs. Smoke must submit to launch needs and clean-air facilities housing satellites as well as to regional air quality guidelines. There is even a weird wildland-urban interface variant in the form of a wildland-space launch complex.

What is striking is that every concern points the same way, to less fire and less intervention. It argues to eliminate fuelbreaks or at least subordinate them to ecological considerations, which lessens the capacity to reintroduce patch burns. It points, in fact, the same direction as virtually everything else in Southern California does, to a schizoid program that defers to wildfire to do the biological work while ratcheting the fire program, click by click, toward suppression. For years the fire program wanted to burn off a dense thicket of manzanita along Bear Creek. One objection after another checked that ambition, the latest being a survey of unexploded ordnance. Then a wildfire in 2010 did the job. That, in cameo, is the regional pattern, disguised at Vandenberg by its quirky security, brazen SLCs, exploding Titans and Minutemen, and such ignition exotica as the filming of the biopic *Amelia*, in which a staged engine fire managed to escape and crash through 700 acres.

————————

In 1896, as the National Academy of Sciences' Committee on Forests gathered to make the recommendations that would lead to an organic act for the forest reserves, nine years before the administration of those reserves went to the Forest Service, the *Los Angeles Times* voiced general alarm over "mountain fires" and urged that the military be called in to control them. As a model it nodded toward Yellowstone National Park, in which Troop M of the U.S. Cavalry had put wandering tourists into reservations called campgrounds and successfully patrolled against fires. Other cavalry units guarded the scenic treasures of Yosemite and Sequoia.[3]

A century later it was clear that military discipline was by itself insufficient to battle fires; much less was martial law a means to govern landscapes. The military had to accommodate fire, and that accommodation would take on the coloration of its context. It would look like fire management in its larger setting. At Eglin AFB this meant a commitment to prescribed fire, a program given afterburners by DOD funding. At Vandenberg AFB it meant fire protection, which led to hotshots, Type III engines, and a willingness to call in air strikes, even with DC-10 air tankers, for initial attack.

America's Gateway to Space may rely on explosive combustion to leave the Earth, but it does all it can to squelch that combustion outside its launch pads. The Gateway passes through a historical portal called Honda Canyon.

AIRING DIFFERENCES

I N THE LOS ANGELES BASIN the contrast between confined and open
burning is awkward because all combustion is collectively trapped
within the barrier mountains and a chronic inversion. The airshed
doesn't distinguish between carbon monoxide released by burning
chamise and carbon monoxide derived from burning briquettes or that
from an automobile's tailpipe. Smog derives from gasoline, but bad air
has many founts. The sources matter less than the common sink.

In 1947 air quality in the South Coast was so awful that public concern
led to the nation's first regulatory agency devoted to a solution, the Los
Angeles County Air Pollution Control District. The consensus villain—
the endlessly useful serpent in the garden—was the automobile, and not
only the cars but the petroleum industry that manufactured the fuels they
required. Worse, until a phaseout began in 1973, the fuel was leaded, such
that lead got deposited on the shrubby hillsides and then entrained again
by wildfires (much as has happened with Ce-182 at Chernobyl). As more
and more of the economy converted to fossil biomass and internal combustion, other points of hydrocarbon emissions emerged and spread like
a contagious fungus. As two memoirists put it, "To us, Orange County
got its name from the color of its atmosphere, not its indigenous fruit."
In 1976 the South Coast Air Quality Management District (SCAQMD)
was created to oversee a comprehensive reduction in contaminants.[1]

By 2010 SCAQMD was regulating 123 stationary sources of pollution, while leaving to state overseers the worst source of smog, the
auto. The SCAQMD regulates gas spouts, furniture solvents, and even

barbeque lighter fluids (it certifies the legal ones). It has programs to replace two-cycle internal-combustion lawnmowers with electrics. It has replaced diesel-powered school buses with compressed natural gas ones. It oversees a car-scrapping program. It restricts open burning by landowners to the combustion of tumbleweeds (I'm not making this up). The quest for clean air extends into the built environment as well. In California it is illegal to smoke in bars or outside public-entry doors.

Yet on the mountains fires burn openly. If fire chiefs had their wish, they would burn extensively for fuel reduction, and if fire ecologists had their druthers, they would burn less effusively and for biological benefits. If fire were restored to the circumstances it knew in 1848 when the Treaty of Guadalupe Hidalgo ceded California to the United States, the South Coast would be routinely smoked in from the endless cooking, ranching, and agricultural burning that went on, along with major smoke episodes when the mountains burned. If researchers are correct that fires lingered for months on the mountains, then the basin would have known a chronic pall for most of the dry months. Paradoxically, the best visibility might occur when Santa Ana winds blow the crud away except for its downwind smoke stream which would have the consistency of a debris flow. There is no easy way to reconcile open burning with a closed basin.

In Southern California every trump card leads to fire exclusion. It isn't enough that fire on the ground threatens houses, thoroughfares, and infrastructure. Even the smoke is toxic, and unless the winds are stiff enough to blow it to the Pacific, it will fill the basin to the brim. Big fires threaten with flame, lingering fires with smoke. Some 98 percent of open burning requests emanate from agricultural fields in the Coahuilla Valley.

SCAQMD insists that it has never denied a legitimate request for prescribed fire on the public lands. It doesn't have to. So sensitized is the public to bad air that the agencies have internalized those concerns and self-censor. They will reach for alternative methods to dampen fuel, or let the uncontrollable wildfires do the ecological work for them. Either way they escape culpability. They did not set a fire that broke loose or clogged the basin with particulates that would choke a diesel semi. At a time when the federal agencies are recommending that employees buy personal liability insurance, it is a daring fire officer who would light up in the name of hazard reduction or ecological servicing. The South Coast Air Quality Management District doesn't have to shut down burning: the 17 million people living within its jurisdiction have chosen to live in a place and in ways that put internal combustion and open flame into direct competition, and there are a lot more people with cars than with torches. The economy runs by burning fossil biomass, not agricultural refuse. You can't see ozone. You can see smoke.

In perverse ways, air quality provides an opportunity to justify not doing what most prudent fire officers consider impractical or reckless. They are better off burning excess wildland fuels in air-curtain incinerators than in situ. The system rewards them for fighting wildfire rather than setting controlled ones. It's a bizarre economy of fire that—in California tradition—lets cataclysms run the market.

As long as the mountains remain vegetated and the South Coast fills with houses and commerce, the basin's airshed will flood with the effluent of combustion, be it smog or smoke. There is no simple solution: probably the number of ideal burn days equals that of extreme fire days. There appears to be no middle ground in a realm where winds slosh back and forth from sea and desert, against and occasionally over a giant granite berm, and inhabitants demand clean air.

Despite the relentless increase in human population and in the population of machines burning petroleum, air quality has paradoxically improved. The reason is largely technological: the evolution of machines has supplanted polluting sources with cleaner ones. In the late 1950s some 1.5 million residential incinerators (along with 17,500 attached to businesses and apartments) fouled the air with combusting garbage; today vehicles haul them to landfills. For similar reasons, replacing old autos with newer ones will likely do more (and can be done) than reducing the number of autos on the road. Technological selection, assisted by the visible hand of the SCAQMD, is casting the inefficient aside.

The sticking point is that there is no equivalent technological solution to wildland fire. The ecology of fire requires burning on site, and it regards smoke as a biological asset, not a contaminant. There is no ecological equivalent of a catalytic converter to segregate combustion from noxious effluents. While the mountains and the basin are both constructed environments, they differ in their relation to fire. The city needs power, not necessarily combustion; the mountains need flame.

The idea—almost comically Californian in its utopian scope—that prescribed fire on public land and free-roaming natural fires in legal wilderness might scale up to something like their former dimensions belongs with Hollywood hype. Such fires will exist only in an alternative virtual reality. The postindustrial future of fire in the South Coast will look like the industrial past. Technological improvements will slowly strangle free-burning combustion out of the built landscape. Ecological imperatives will reinstate free-burning flames on wildlands. The difference is that fire's removal will have economic markets and political calculation behind it, while fire's restoration will likely happen by political default and chance. The one will come through engineering sparked by clever ideas, the other by conflagrations kindled by accident, arson, or (rarely) lightning.

The pyric geography of the South Coast will be determined by lines drawn in the dirt. Many will be etched with fuelbreaks as demilitarized zones to segregate the wild from the urban. Under extreme conditions these will fail as sparks fly over and flames slop over the borders. Still, land use will mostly separate the worlds of open flame from those of internal combustion. But no such barrier, however nominal, exists for the air. Particulates from diesels will mix with those from manzanita, carbon dioxide molecules released by lawn mowers will mingle indistinguishably with those from burning ceanothus. This unsavory brew will be palpable to the general public—smelled as well as seen—in ways that flames along the fringes of the San Gabriel Wilderness will not be. The only way to prevent that atmospheric mix-up is to stop the fires from happening.

It's axiomatic that the effluent we can't see is often worse than that we can. So, too, the controls over fire can happen off site, or internally, invisible to public observation. In the complex card game being played out over combustion in the South Coast, air quality is the ace up the sleeve. Until there exists a mechanism to segregate flame from smoke, the differences will sharpen and the greater megalopolis and its environs will become more combustion unstable. Eventually, the South Coast's autos may be powered by electricity generated from wind turbines located outside the basin while its biota burns by conflagration. The visible hand that once held the torch will be replaced by an invisible one adjusting monitors to gauge greenhouse gases.

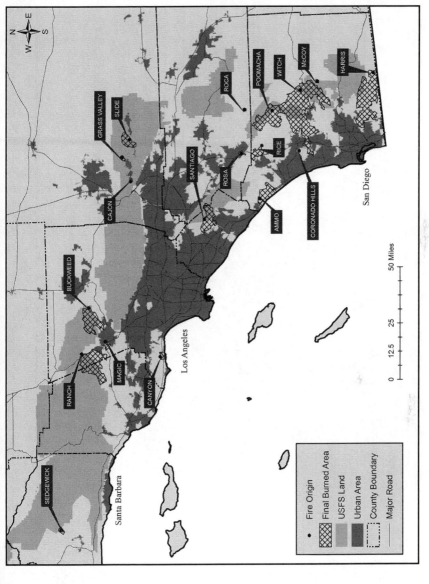

FIGURE 1.
California—two
fire provinces,
two ecologies,
two dynamics of
burning, but a
similar outcome.
(a) Southern
California fire
siege of 2007.

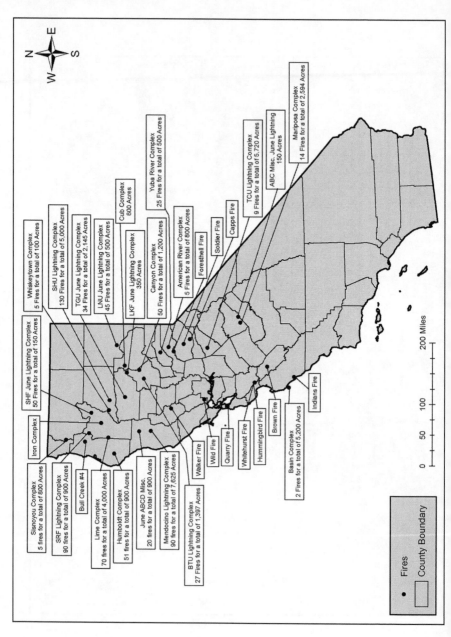

(b) Northern California fire siege of 2008. Original maps by CalFire, redrawn by Chris Miller.

FIGURE 2. The wildland-near Earth interface, Vandenberg Air Force Base. Photo by author.

FIGURE 3. Inaja fire fatality scene. Photo by author.

FIGURE 4. Angora fire: the threat and the promise of fire at Lake Tahoe. Photo by author.

FIGURE 5. The other fire scars: Oakland Tunnel fire memorial. Photo by author.

FIGURE 6. Fire and the Big Trees. Fire scarring abounds, which it should since fire and the sequoias have coevolved. Photo by author.

CALIFORNIA SPLIT

THE TWO CALIFORNIAS MEET, often talk, and sometimes share and socialize, but they don't freely mix. They are yoked out of necessity, not desire, and through human institutions, not natural logic. Only during emergencies is there a semblance of commonality when suppression compels a working unity. Outside those cataclysms north and south have their characteristic species, their characteristic fire outbursts, even their characteristic career paths.

Yet the postwar drive for a statewide master fire plan mandated that whichever half was dominant would impose its paradigm on the other. In the early years of formal fire programs Northern California challenged, and defeated, light-burning as a foundational doctrine and replaced it with systematic fire protection, which became the basis for planning in national forests. Fifty years later, during the years of the fire revolution, Southern California challenged, and redirected, fire restoration as a goal and replaced it with an urban fire services model. The upshot is that California as a state has repeatedly pushed toward fire's exclusion.

Still, there were pockets of resistance—California was too big and too diverse not to have countercultural nooks. Light-burners were the bohemians and beatniks of their day, fire restorationists were the back-to-nature communes of theirs. Some remained symbolic—both north and south had iconic species; some were large enough or quirky enough to claim national attention, like new cults out of Big Sur. The big burns, and each region had its signature variety, drove the system, and because it was California both the quirky and the catastrophic assumed exaggerated standing in the American fire community generally. By the 21st century the divide was less between north and south than between wild and urban.

CHARISMATIC METAFLORA

North and south can differ within themselves as much as between each other. No single regime defines either, which makes it even more preposterous to anoint a single species to characterize the whole. Yet, in the way of symbolism, each region has its honored pyrophyte. They are the poster children for what fire means and how it ought to be managed in each domain. They quickly escalated from the status of scientific sample to value-laden metaflora. For the south, it is chamise (*Adenostoma fasciculatum*). For the north, it is the sequoia (*Sequoiadendron giganteum*). Each emerged in the public limelight during the 1960s, and they represent the inchoate split in fire philosophies that were then struggling into definition.

Chamise acquired celebrity, if not charisma, when it was adopted by the Riverside Fire Lab as the model species for the regional fuel complex. Its association with fire made chamise something more than another spiky-branched member of the unloved "brush." It was the yin to the Santa Ana's yang that together powered the remorseless conflagrations that threatened the South Coast.

The behavior of large fires underwrote much of the research agenda of the lab. Interest stemmed not only from catastrophic fires that blasted Southern California from 1955 to 1961, but from national concerns over a possible nuclear war with the Soviet Union after the 1962 Cuban missile crisis. Region and nation, Forest Service and Civil Defense, scientist and practitioner, all were desperate to understand the mechanics of big fires, or what was being termed mass fire. Project Flambeau (1964–67) sought to understand how such conflagrations start and spread. Simultaneously, extensive research into fuelbreaks sought to learn how to slow or retard them. Southern California provided an ongoing experiment: no one needed thermonuclear warheads to kindle field demonstrations of mass fires that could engulf mountains and suburbs. What fueled those fires was chaparral. And a primary constituent—what could serve as shorthand for the whole complex of combustibles—was chamise.[1]

That was why chamise mattered. But of all the properties of this complicated plant what mattered to research were those traits that allowed it to burn. What they measured were those physical and chemical parameters that could feed into the existing fire behavior model being developed at the Missoula Lab for the National Fire Danger Rating System. They wanted to quantify the distribution of fuel in an array by size, condition (living or dead), loading, density, surface-to-volume ratio, fuel bed porosity, moisture content, heat content, and chemical composition. The biology of chamise was relevant to the extent that it shaped seasonal changes in moisture and volatiles and secular changes in the physical properties. Eerily, as chamise passed through its annual phenological cycle, the tendency to burn improved in sync with drought and winds, reaching a maximum during late fall Santa Ana conditions. As it aged it transformed more of its biomass into available fuel, such that the longer it remained unburned the more susceptible to burning it became. Before five years it hardly burned at all; after 25, it burned with increasing intensity. And once burned, it reseeded prolifically in the ash or resprouted from its buried stem. It seemed, in fact, a perfect phoenix species whose life cycle readied itself for immolation.

Chamise, in brief, did not thrive despite fire but seemingly because of it. The plant grew fuel, the fuel burned, the fire renewed the plant, which quickly resprouted and reseeded, so it could refuel the landscape. The system was almost mechanical in its regularity, like the piston in an automobile that sucked in combustibles and oxygen and then compressed them until the spark of ignition exploded the mix. Because it was impossible in Southern California to rebuild the mountains, calm the winds, or prevent ignitions by accident or arson, the only way to shatter the cycle of conflagrations was to target the chaparral. The life cycle of chamise showed how and when to do that and why the strategy would work.

The outcome took many forms, some based on the science, some not, and some—like the worst of the fuelbreaks on the Angeles after the 1970 season—best considered political graffiti. But fuelbreaks elsewhere, and brush conversion schemes that alloyed with wildlife habitat, range improvement, and watershed enhancement, became the treatment of choice throughout the region. It reduced a lot of quirky biology to codes fit for the prevailing physical models of fire spread. It seemed to align with a revival of interest in prescribed burning nationally. It was for California what the understory rough was for Florida. Fuel management was the future not only of fire control but of fire management overall.

This was a lot of ask of one shrub. Inevitably perhaps the research was extrapolated far beyond the range of its meticulous measurements, much as happened with the Rothermel model that identified the parameters to be recorded. Originally devised for the National Fire Danger Rating System, the fire behavior model was seized upon by practitioners and administrators and flung far beyond the boundaries of its experimental scope. The need was urgent. Fire officers used whatever was at hand. If it came with a scientific imprimatur, so much the better.

Ideas, too, have their life cycles, and they can senesce and eventually burn away. The ardor passed. Chaparral was more than fuel, prescribed fire involved more than hazard reduction, fire management meant more than aggressive fire suppression. Further research argued for the complexity of the fire environment and the limited value of wholesale conversions. The most meaningful fuel reduction pointed to the suburbs, not the brush. In the end, chamise did not so much inform strategy as the push toward suppression informed the science. What began as highly targeted research became synecdoche for a larger strategy. It endures as a symbol, not so much of chaparral as of the way everything in Southern California, nature's economy no less than people's, impels toward a binge-bust cycle of conflagration and control. In the end the only strategy allowed was suppression.

The sequoias propose a very different story, although that progressed in uncanny lockstep with chamise. The critical facts were that they were already charismatic and they would contribute only in minor ways to the fuel array of their habitat. The Big Trees had deep taproots in the American pantheon of sacred places: three of the first four nationally authorized parks were established to protect them. Yellowstone might celebrate a big-sky wilderness, but Yosemite, Sequoia, and General Grant National Parks were originally founded to protect the cathedral groves. Similarly, the sequoias' role in fire ecology did not bring them glamour. Rather, their transcendence bestowed grace on fire ecology. They made fire management something more than the cycling of combustibles. They became the poster child for a fire science grounded in biology, not physics.

The public had long identified sequoias with fire because the evidence was so stark. They could see the enormous cavities burned into the trunks, and the earliest settlers reported near-annual surface fires that washed through the groves like spring melt. Early administrators brooded over whether to control-burn as a sanitizing measure. But it was direct human impact by simple trampling that catalyzed the modern era. The early studies spanned from 1956 to 1962, and after the Leopold Report of 1963 argued, using the Sierra Nevada parks as exemplars, for a rechartering of National Park Service administration in natural areas, researchers pivoted toward fire's effects, particularly its impact on reproduction. With inspiration from the Leopold Report, not the Cuban missile crisis; with sponsorship from the National Park Service, not the Forest Service and Department of Defense; with researchers drawn from the Bay Area, not Southern California; and with much of the research conducted on the University of California, Berkeley's Whitaker's Forest, not the San Dimas watershed, it should come as no surprise that fire in the Big Trees meant something very different from fire in the brush.

The early research by Richard Hartesveldt broadened to include botanists, entomologists, zoologists, ecologists—a full-service survey of sequoia habitat over 1963 to 1970. The focus was on the survival of these megafloral relics, especially their mode of reproduction. That was where fire fit in. It triggered the semiserotinous cones, cleared the land to receive seeds, helped thin thickets of young saplings, and fended off competitors by burning away invading fir, cedar, and pine. Such fires left the asbestos-barked and high-branched sequoias untouched; or where debris at the base allowed fire to burn into the trunk, perhaps excavating a cavity, the trees shrugged off the trauma as a pyric bruise. While fire was only one factor among many in regeneration, it was vivid, and essential. Without proper fire management, the Big Trees would die out much as Shasta ground sloths had and California condors would if not carefully nurtured.[2]

The concern, that is, was with the propagation of the species, not the propagation of flame. Researchers fretted over "fuels" because decades of fire protection had allowed debris to collect around and within the great groves, and mixed conifers, notably white fir, had silted in the pore space between groves. They deposited thick layers of duff and humus that prevented any sequoia regeneration. And if those fetid woods burned, the flames could threaten the grizzly giants themselves in ways never possible in the past.

Such fuel buildups were unnatural, a product of fire's attempted exclusion, not a part of the natural order. The sequoia did not fuel fire: it seized on the opportunities fire presented. To control-burn or crush chamise short-circuited the fuels cycle but also renewed it; the intervention did not alter the fundamental dynamics. Doing the same in sequoia groves could remove the offending fuel invasives and restore the scene to a more primitive condition. The chamise made fire possible. The sequoia was made possible by fire.

Each characterized a very different regimen of fire and of fire's management. Chamise suggested an endless and inevitable cycle of conflagration, powered by the shrub's very existence, for which fuel had to be removed or fires fought. Sequoias suggested a world of friendly flame that regularly swept the cathedral groves clear of quotidian clutter and made possible a renewal of the spirit. The former subordinated land management to fire management. The latter placed fire well under the command of the Generals Grant and Sherman trees.

FIRE BUSTS

Big fires, too, divide along the fault border, and this being California, their contrast is exaggerated, and their character archetypal. Southern California fires have the swagger and hype of Hollywood action heroes. They crash into the scene like Godzilla or an army of invading aliens. They overpower control efforts by sheer momentum and heft. Northern California fires come like a flash flood or a sudden, crippling blizzard. They appear like a cloud of lightning-buzzing locusts, swarming over the land and defying control by their overwhelming numbers.

They've both haunted the scene in recent times, storming over the land with such overawing power that they are no longer even called busts but sieges. The institution-shaping events of the modern era date back to 1970 and 1977. Then in late August 1987 lightning pummeled Northern California, kindling 4,161 fires (92 of them over 300 acres) that burned across 755,475 acres, consumed 42 houses, and killed 10 firefighters. Smoke clotted the Sacramento Valley for weeks. It was California's worst year for burned acreage since modern records accrued in the 1930s—and the origin of the term *siege* for this kind of fire bust. (Previously, large firefights had been characterized as *campaigns*, in which the agencies took the fight to the flames. Now, they found themselves bogged down in siege warfare, or perhaps themselves besieged.) The early 1980s had been slow seasons for fires nationally; the Siege of '87 forced the fire establishment to relearn how to mobilize on a huge scale, preparing it (and the media) for Yellowstone's summer of fire the next year. Southern California returned to the stage in 1993. The Malibu drama had been played out many times before, but like a Hollywood remake of a silent-era film, this time it was performed in Technicolor with a cast of thousands.[3]

What the busts shared was a capacity to overwhelm any initial attack that firefighting agencies could muster. In 1987 there was no way enough crews, aircraft, and engines could be assembled and sent out through the backcountry in time to catch every fire start in the first burning period. But so vast were the number of ignitions that even if only 2 percent escaped it meant scores of fires scattered across hundreds of square miles in often lightly roaded countryside. Until the weather broke, there was nothing to do but pull back, burn out, and wait for the skies to clear. The fires burned—and smoked—on and on. They burned through the Labor Day weekend, they burned through all of September, they didn't cease until October 3. They were sustained by an immense high pressure cell that stagnated over the region, heating the land, drying biotas, and capping the atmosphere. The valley smoked in like a cabin with an open fireplace and a closed flue.

In 1993 it was Southern California's turn, as arson, accident, and Santa Anas sent fires pouring over the mountains like outflow from a ruptured dam. Even a smoothly oiled machine like the state fire plan could not assemble force enough fast enough to halt an epidemic of ignitions fanned by Santa Ana winds. The fires simply blew past or over them. What slipped past initial responders became big. Again, it was not just the fact that there were not enough suppression forces to cover all the assets because it was the fringing houses, not the interior acres, that most absorbed resources. The issue was that agencies could not rally as quickly as the wind blew. The final tally helped confirm what was becoming a new norm: 21 fires, 197,225 acres burned, 1,241 structures lost, and three dead. What moved the episode into apologue was the Breughelian spectacle of celebrity Malibu in disarray before the well-filmed flames.[4]

In the new millennium the southern scenario repeated twice more. In 2003 an outbreak ringed the mountains—a fire siege, indeed—and loosed a phalanx of major fires across 750,043 acres, incinerated 3,710 houses, and killed 24 persons. It was nature's version of "shock and awe." Two such fires at once would probably overrun the most hard-bitten system. Southern California had 14. Under the state fire plan some 1,572 engines, 136 aircraft, and 14,000 personnel worked the firelines. It didn't seem possible to mutate further, or burn more or with still greater damages. And the 2007 siege was marginally less horrific, as some 16 major fires slicked off 518,000 acres, turned 3,069 houses to ash, and killed 17 people. Nearly half a million residents were evacuated. It was the fire equivalent of an earthquake or tsunami. Recovery, however, followed fire norms, not those of deeper shaking and baking. Still, the outbreak suggested, with hideous indifference, that Southern California might be facing a new future of firefights.[5]

But as the South Coast readied for its next round, the big fire scene headed north. In June 2008 what dazed commentators labeled a freakish storm kindled 1,754 fires from Big Sur to Yreka. There were too many starts to contain with initial attack. Amid the driest spring in 114 years some kindlings grew large; many burned together; individual "fires" merged into sprawling "complexes." The ordeal lasted between three and five months. When it ended, 1.2 million acres had burned—a new record—along with 186 houses. Against the flames the agencies had mustered some 20,000 firefighters. Thirteen people died. By area the Siege of 2008 accounted for six of the 26 largest fires recorded in California since 1932.[6]

The real surprise is that anyone should continue to be surprised. The lightning storms were not freakish, the Santa Anas not abnormal: this is how fire in California behaves, a chronic backdrop of tremors and storms, punctuated every decade or so by shattering eruptions. The fires do not return year after year like winters in the Midwest. California burning is not a bell-shaped curve. The big years do the major ecological work, cause the greatest damage, and ring up the most exorbitant bills. Fire protection means managing for the Big One, even if the Big One comes in a baker's dozen. That is true both north and south.

Still, their peculiar characteristics grant each region a fire personality. Fire busts in the north are widely ranging, sullen, and tenacious. Those in the south are fixed to the mountains, mocking, and explosive. That they find themselves linked is not due to any intrinsic logic of geography or burning. Like pyric plates, they grind against each other in the tectonics of California combustion. What joins them is a collective institution that can ship engines from north to south and hotshot crews from south to north. Not wind but aircraft, not fuels but dozers and handcrews lash their fates together. They are joined not by mutually shared fire regimes but by programs of mutual aid to cope with conflagrations. The needs of the south drive those of the north.

ARCH ROCK

PORTAL TO NORTHERN CALIFORNIA

THE WENDING ROAD, like the Merced River it runs alongside, rises from the Central Valley into the Sierra Nevada. But the land to its flanks rises faster. Rounded hills steepen, the valley narrows. Grass and oak become blotchy with chaparral. Loose shale and brush yield to patches of slicked granite. Beyond El Portal the granite patches thicken into sheets, while a riparian gallery forest merges into valley-veneering conifers. Beyond Arch Rock the gorge suddenly widens. The granitic sheets panel the sides with walls, spires, and domes. Water thunders over hanging valleys, and in the spring it trickles down slickrock like lace. Yosemite Valley suddenly fills the panorama.

The abruptness of the vision is part of its aesthetic; perhaps only Grand Canyon, invisible until one stands on the rim, can match the visual impact of that unanticipated revelation. The scene seems like another world: it radiates the sense of a place preserved fresh from the Creation, a sanctuary from quotidian American life. Entering Yosemite Valley is like passing through a portal to a prelapsarian past.

The portal is a geologic testimony to water. Ancient water, frozen into the glaciers that sculpted the valley. Modern water, boiling down the Merced and storming over the waterfalls that form a rampart around the valley's rim. Contested water, forcing politics to decide what to leave free flowing and what to dam. Add a storm, and the sense of a watery and watered world, lush with greenery, fills the granitic frame. But the scene could as easily be defined by fire, and not just the metaphoric fires that sent granite plutons bubbling upward in eons past.

On August 7, 1990, lightning kindled two blazes in the Arch Rock region, one to each side of the Merced River. They burned savagely for almost two weeks. The Steamboat fire scorched 5,325 acres and the A-Rock fire some 18,100 acres. They shut down access, like the flaming swords of the cherubim that guarded Eden. Yet they were a biotic reminder, an ecological premonition, that fire would trickle through the backcountry and occasionally pour across it as fully as streams and waterfalls. The Sierras burn. They are the second axis to California fire, a tectonic Y to the Transverse Range's X.

What moves that fact from natural history to moral drama is that Yosemite National Park, along with its Sierra sibling, Sequoia-Kings Canyon, serves as a sanctuary for natural fire. They shield a small but far from trivial fraction of the Sierras from the remorseless pressures of fire suppression. In the parks fire officers are still called managers, not chiefs. Here fire serves the agency, not the agency fire.

What moves such facts from anecdote to political narrative is that the Sierra parks have succeeded in projecting their ambitions and influence onto a national screen. They were the proof-of-concept test for restoring natural fire.

=====

From its origins Yosemite has claimed iconic status in the history of American environmentalism. The valley was discovered by fur trapper Joseph Walker in 1833 and then rediscovered by prospectors in 1851. But its mineral wealth offered a scenic rather than a commercial mother lode.[1]

In 1863 Albert Bierstadt painted the valley and placed it within the pantheon of heroic places. A year later the federal government created within Yosemite the country's first national nature preserve, although it later ceded control to the State of California. At issue was not the scene, for the valley was a granite citadel seemingly immune to anything humans might do. What mattered was the sequoia groves that dappled across the Sierra front. The Mariposa Grove was vulnerable to logging and, so it appeared to early parties, to fire. They reported that the grove was burned almost annually by the indigenes. Yosemite thus became the nation's first protected park. The move to shield it concentrated on the Big Trees and the threat posed by axe and fire. The Mariposa became, for secular America, its sacred grove.

The park returned to the federal government in 1890 and soon expanded. The great prophet of wilderness, John Muir, made it his special habitat. He, too, railed against wanton fire; fire, he reckoned, destroyed 10 times as much forest as the axe did, and, as bad, it was the means by which sheepherders with their detested flocks, which he famously denounced as "hooved locusts," readied their high pastures. In 1903 Muir spent a night in the valley with President Theodore Roosevelt, arguing the case for conservation. They had their picture taken at the base of a giant sequoia; that night it snowed on them. In 1913 the valley's granite double, Hetch Hetchy, became the emblem of a martyred landscape when legislation authorized it to be dammed to supply water for a San Francisco still struggling to recover from the 1906 earthquake and fire. Later Ansel Adams affirmed Yosemite as the definition of wilderness aesthetics. The Sierra Club remade environmental activism. The valley's Camp 4 climbing culture transformed wilderness recreation. Over and again the informing issues of America's environmental movement were fought out at Yosemite and, with somewhat less celebrity, at Sequoia-Kings.

———

That included fire. Over the course of the 20th century Americans had spent the first 50 years struggling to take fire out, and the next 50 straining to put some of it back in; both campaigns were initiated in the national parks. The U.S. Cavalry that oversaw the parks until 1916 established the paramilitary paradigm of fire suppression that the Forest Service later perfected. Then, in the 1960s, the Sierra parks became the paradigm of progressive fire management. While the Leopold Report (1963) began with the question of elk in Yellowstone, it soon relocated to where the real action was—the Sierra Nevada.

Yosemite became the poster child for the unintended havoc wrought by fire exclusion. The valley was so overgrown with woods that it was difficult to see its fabled walls and waterfalls. And the sacred Mariposa Grove was overrun with a profane invasion of white fir that promised to submerge it like coniferous kudzu. Banning all fires ultimately threatened the grove with horrific fires and condemned the prospect for sequoia regeneration. The Sierra scene presented an antithesis to the Southern California model, and established a third apex to the national fire management triangle. It became for the promotion of natural fire what Florida was for prescribed fire. With crisp, unmistakable symbolism, in 1968 Yosemite ended the tourist-delighting Firefall from Glacier Point and replaced it with let-burns in the Illilouette Valley. The availability of remote, quasi-isolated basins in the high country simplified a program committed to natural fire.

The Sierra parks became a protected enclave outside the California fire master plan. They could not reform the California system any more than national parks can reform American society; but they could endure as sanctuaries able to resist having their mission suborned to the demands of emergencies and they could advertise an alternative vision of fire, for which they developed a parallel program. The acres involved were not many, but they loomed large because they defied collective wisdom that fire could not safely be restored. If Southern California stood for wildlands under the administration of urban fire services, the Sierra parks proposed to administer fire management under a commitment to wilderness or its "reasonable illusion."

Passing through that Merced portal is like passing through a looking glass in which fire agencies promoted fires rather than attacked them, wildfires became opportunities rather than threats, and the fire service took its inspiration not from the all-risk emergencies of contemporary life but from a past world in which people had not yet trammeled the land and could experience soaring flames with the same awe they viewed falling water. The world beyond was not large—most of the Sierra Nevada and nearly all of northern California lay outside it. But California had always found niches for the asocial and the anomalous. The Sierra parks were the old Bohemian Club's camp in the sequoia groves writ large.

THE TAHOE CRUCIBLE

L AKE TAHOE would seem to have it all.
It's a spectacular natural phenomenon—North America's answer to
Lake Baikal. Within the Sierra Nevada, itself iconic for American
environmentalism, only Yosemite Valley rivals it for geologic charisma.
With astounding clarity it forms a naturally bounded entity, a contained
basin almost utopian in the purity of its configuration; and with astonishing fidelity that catchment has collected the history, ecology, economy,
politics, and character of the American West. All the pieces are here, and
although they have assembled themselves in unique and stunning ways,
the story is a familiar one. Fifty-three watersheds drain into the lake, only
one, the Truckee River, leaves, and it tracks the path of regional history.
Lake Tahoe is a microcosm. If it is possible anywhere to distill the northern California fire story, this is the landscape to do it.

That hope works: the past is present. What is disturbing, however, is
what Lake Tahoe means for the future.[1]

═══════════

Despite its lofty perch along the Sierra summit, Lake Tahoe has never
been isolated. The Washoes trekked to the basin each summer. One of
the main routes into California for emigrants, over Donner Pass, brushes
its northern rim. Placer miners from the golden age sluiced up from the
mother lode country, and then hard rock miners dug into the Sierras over

its eastern rim at Virginia City, where the Comstock defined a silver age. The transcontinental railroad followed the Truckee River over the mountains. The lake was already a tourist destination when Mark Twain visited it in 1870. Loggers built railroads and cut out the Jeffrey pine by the end of the 19th century. When they left, pastoralists moved into the stump fields, cattle mostly around the wetlands and near-lake environs, sheep up the slopes. Land changed hands among big owners. Then the conservation movement pointed to the wreckage and surrounded the lake with national forests on what remained of the public domain. As early as 1883, two years before it created a state Board of Forestry, California established the Lake Bigler Forestry Commission to inquire into means to salvage the scene; in 1913 the Lake Tahoe Protection Association organized the environmental enthusiasts of the day.[2]

It's a common scenario, replayed throughout the West with local accents, and it has yielded a common chronicle of fire. Lightning kindled fires in the basin, although fewer than in the surrounding landscapes and, more quizzically, out of sync with the great lightning busts that from time to time shake and bake Northern California. The Washoe added their own burns; fires along the lakeshore or low meadows would have crept and swept upslope during critical fire weather, and the southwest winds could have pushed flames over the eastern flanks. Fire scar records suggest a return interval of 8 to 10 years, typical of midlevel conifer forests. With the advent of American settlement, the tempo quickened. More fires, more slash—fires increased in number and severity. The Herodotus of forest reserves, John Leiberg, documented the successive fires that were the camp followers of miners, loggers, railroaders, and shepherds. Altogether he reckoned that "extensive fires have swept" all the tracts surrounding the lake, although few involved "total destruction" and most thinned and scabbed the land into brush fields.[3]

That helped argue for protection, which came in the form of forest reserves and, with the U.S. Forest Service, effective administration. Gradually, logging quieted, not least from having cut over the mature forest. Herding shrank. Fires died down. A new forest grew up, but a different one, with few Jeffrey pine, a greater mixture of species, and a far denser congestion of combustibles, an assemblage more likely to move fire from the surface to the crowns. A vigorous firefighting effort and the removal of traditional burners held the threat in check. The CCC added muscle;

it planted its first camp in 1933. Since the lake occupies most of the basin, fire protection had only to deal with the donut of combustibles between the lakeshore and the granite summits. Fire spread had to sandwich between them. A fire anywhere could readily be seen from everywhere.

Then came the postwar boom. Since the basin had already been stripped of its commodities, and because it enjoyed good transportation and proximity to urban centers, it made the transition to a consumer culture and service economy quickly. The Nevada shore replaced mines with casinos; clear-cuts yielded to ski runs; tourists substituted for sheep; governments replaced robber barons as large-estate landowners. Recreation became the driver of commerce. Workmen's shacks morphed into summer homes. Sportsmen fished for trophies rather than bulk stock. In 1960 Squaw Valley hosted the Winter Olympics. By now, campfires and construction equipment started most wildfires. Few grew beyond a handful of acres.

Lake Tahoe urbanized. Once again, it became a sampler of the western experience. There were clusters of cabins, relic log mansions, crossroads hamlets, and gnawing out from the shore recreational development. By 1970 the Tahoe basin had 49,000 subdivided lots and hundreds of miles of roads to them, the modern avatar of homesteading. In tribute to the postwar car culture it also devised a world-class strip mall. Unlike many western metropolises Lake Tahoe does not have an exurban edge city: it has *only* an edge city. Ringing long stretches of the shoreline is a skirt of casinos, restaurants, small-shop chalets, resorts, cramped motels, apartments, trophy homes, Irish pubs, sporting goods outlets, Safeways and Staples, and the urban paraphernalia, including an artificial marina and airport, that a resident population of perhaps 60,000 and a seasonal influx of several million tourists demand. If uncontained the growth would ring the lakeshore and turn on itself like a Möbius strip, if a Möbius strip could be a strip mall. Observers note that its economy is no longer based on commodities but on experiences.

With that shift came also the advent of modern environmentalism. If the Sierra parks were vignettes of primitive America, Lake Tahoe was a vignette of postindustrial America. Instead of the primitive, it offered the pristine. When its waters turned murky from effluent and its skies smoked in, it became a postcard from the dark side of the Sierras. The character of the lake and basin meant it was easy to identify contaminants and their sources. The dazzling purity of its waters defined its state of nature. There was nowhere to hide pollution from sewage, construction runoffs, or eutrophication. What happened in Tahoe would not be lost or hidden in a remote backcountry. This was a celebrity landscape, to California what Everglades is to Florida, and among the most scientifically scrutinized places in America. It's as overrun with transhumant tourists as formerly with sheep and as picked over by researchers as by an earlier generation of prospectors. Its snowy ramparts are visible from the capitals of two states.

Restoration was a matter of self-interest as much as of public ones, of economics as well as environmental ethics. In 1969 California and Nevada signed the Bi-State Compact to create a Tahoe Regional Planning Agency to bring some coherence to development (and avoid a ruinous competition). But the complexity of diffused responsibilities made decisions tricky and enforcement almost impossible. In 1982 the Tahoe Regional Planning Agency proposed limits on what kinds of use the basin should tolerate; five years later it enacted a code of ordinance that prescribed where and how future development might occur. Almost 20 years had passed. More than goodwill and legal proclamations catalyzed reform. Behind it lay a serious program of land purchases by both states and the federal government in which acquisitions targeted large estates along the lakeshore that otherwise would have been subdivided. Public ownership stepped upward from roughly half the basin to 85 percent by 2000. That transfer stabilized development and granted leverage to public oversight.

The simplest solution would be to submit to the geography of the basin and create a single governing entity. The campaign for national park designation began in the 19th century, but every time the issue revived, it stalled against the same critique. The parks were preserves, holding inviolate vestiges of primitive America, while Lake Tahoe had been degraded, not once but repeatedly, by every commercial ambition that blew like the Washoe zephyr over the West. Miners had ripped open rock, loggers had stripped nearly all its forests, pastoralists had clipped its meadows and grassy understories, recreationists had planted summer homes and ski resorts—the idea of a national park did not, at this time, include landscape-scale restoration. Every feature that spoke to Tahoe's natural wonders had made it vulnerable to human finagling. The magnitude of Tahoe's ravished past precluded it from park status. And of course there was also agency rivalry since the Forest Service, the primary public land-holder, was ever wary of poaching by the Park Service. Yet even under the Forest Service, Tahoe was divided among three national forests, which often struggled to reconcile the basin with their other, larger domains and then among their own competing interests.

The solution was to fashion a park equivalent in 1973 by consolidating the basin portions into a new entity, Lake Tahoe Basin Management Unit. The management unit was not officially a new national forest any more than it was a park, but it functioned like one. Still, even though the Forest Service controlled 75 percent of the land, it had to work through a dog-hair thicket of other public agencies, nongovernmental organizations, and private interest groups, not least the basin's division between two states, one with legalized casinos and one without. Multiple use edges from paradox into parody, as 30 ski resorts rub shoulders with three wilderness areas that spill over the rim and a circuit of roadside commerce clings to the shore. If the built landscape were as integrated as the natural, the place might stand as America's most spectacular theme park. Instead, ill-constrained development elevates tackiness to an Olympic sport. At Tahoe nature proposes one unity, and people labor to impose another.

The problems of restoration—of retrofitting modern sensibilities onto landscapes overturned by earlier eras—have proved very different from the problems of protecting a legacy landscape. The Tahoe basin is not so much an ecological island as a historical cauldron into which a witch's brew of western themes has been stirred over various incantations. The latest iteration followed a presidential visit by Bill Clinton and Al Gore in 1997, at the invitation of Nevada senator Harry Reid, which led to funding for a comprehensive summary of environmental issues published by the Forest Service in 2000. The outcome was an impressive, multi-disciplinary survey that reaffirmed Tahoe's status as a "powerful laboratory for learning." Few places have been subjected to such intense scientific inquiry—or to such ongoing public scrutiny.

The organizing theme for the task force study was of course water—the watershed that washed into and drained out of the lake, the fabled purity of the lake's waters. But Tahoe was a fireshed as well. As land use had changed, so have its fires such that a renewed coniferous forest—some 85 percent of the basin's woods are regrowth—has reinstated fire. Exurban and recreational development are embedded within or trim that restored capacity for burning. In a curious way, the basin replicates the clash between wildland and urban, which occurs along their unstable border. That pyrogeography may not seem apparent because an incombustible lake assumes the role of a large metropolis. Big fires, however, don't burn in the core, any more than logs do. They burn along the fringe.

Lake Tahoe miniaturizes the contemporary California scene the way a microchip does a vacuum-tube computer. In a casual way Tahoe is to Northern California what Lake Arrowhead is to Southern, but their differences may matter more than their similarities. Most simply, it's a question of scale. Lake Arrowhead and Big Bear Lake are significant to the region, Lake Tahoe to the nation. At Arrowhead the Mountain Area Safety Taskforce, devoted to treating fuels on a landscape scale, has succeeded because it involves a big investment in a relatively small place. Tahoe has comparable concerns but a more complex political environment and much higher funding needs. But even more, although Tahoe is a large place—you could probably put the San Francisco Peninsula from the city to Silicon Valley into it—it has only small room to maneuver.

What makes the inhabited beltway attractive—its binocular proximity to lake and crestline—also squeezes it between water and rock. The forest that might act as a buffer, and does visually, can become kindling with little prompting. The first literary account of the lake, in Mark Twain's *Roughing It*, famously ends with a conflagration that drives Twain to a boat when his campfire flares out of control. The forest may have exchanged white fir for Jeffrey pine, as the gambling emporia have replaced three-card monte with slots, but the house odds still favor fire.

What makes the Tahoe fire scene special is not its wildland-urban interface—which has become a norm throughout the West—but the lack of space in which to address it. Between its edge cities and its granite rim, there is little geographic play. Few natural barriers on which to make a stand or backfire. Few anchor points or safety zones. Few patches with which to break spread. Virtually no opportunity to use free-burning fire by prescription, not only because of escapes but because of smoke and ashy runoff into the lake. Yet to install an infrastructure for aggressive suppression would be a form of prohibited development in itself. Fire roads and fuelbreaks can contaminate the lake as surely as roads to cabins.

Nor is there much political space in which to maneuver. Public scrutiny is relentless; anything done anywhere in the basin is visible everywhere. Restoring water, not fire, is the ultimate ambition. Almost everything that happens in the lake, whether for development or restoration, includes a fire component, but fire must submit to those other purposes. Wildfire threatens the built environment. Prescribed fire compromises air quality and potentially water quality through runoff. The primary threatened and endangered species—the spotted owl, northern goshawk, willow flycatcher—favor older, dense forests. Landscaping challenges homeowners' desires for privacy and an aesthetic for dark woods and if done rudely can send debris to the lake. A 2006 survey of fire potential concluded that 60 percent of the basin's forests could support crown fire and 76 percent of those in the wildland-urban interface could; some 70 percent of houses lacked defensible space. There is little tolerance for mistakes. Tahoe is, in the words of one fire officer, a "political fishbowl."[4]

Yet a program of fuel reduction goes on, year by painstaking year, community by community. In 2007 a more comprehensive approach matured with the Lake Tahoe Basin Multi-Jurisdictional Fuel Reduction Plan, a 10-year program among 16 agencies to prepare community wildfire protection plans, encourage fire safe councils, and mechanically treat 68,112 acres. Adding urgency to the project a fire that same year ripped across 3,072 acres on the eastern slopes of Angora Peak and burned over 250 structures in something like three hours. That commanded a lot of attention. Among the conclusions of a postfire review was the concept of "crown fire momentum" in which, even in treated areas, a full-blown crown fire can continue some distance before being forced to the surface, where the full effects of thinning can prevent a rekindling in the crown. That further reduced Tahoe's room for maneuvering. But then that is what every factor did. In Florida—say, Everglades, an analogous reconstruction project—every factor could encourage prescribed fire. At Tahoe every consideration shut fire down.[5]

Just as Tahoe gathered and compressed much of the fire story for the West, so it illustrates the difficulties of fire's restoration. It demonstrates, again, that it is much easier to take fire out of a landscape than to put it back in. More broadly, it offers a fire version of the problems of retrofitting. Or more bluntly, it highlights that America is not good at cleaning up after itself. It favors the pioneer, the forty-niner, the first mover. When the burdens of history become onerous, its preferred solution is to be born again, declare bankruptcy, send back the keys for houses with underwater mortgages, reinvent one's self. That's not an option at Tahoe. What is brought to it stays. The bounds of the basin's history are as sharp as those of its geography. Tahoe's environmental crisis was a long time developing; it took decades to strip and refill with sewage, slash, and schlock; and it will take decades to reconstruct it.

The deeper message, however, may be the power of scale. The task of managing fires is easier in the Mogollon Mountains or the Yukon Valley, where the fires are distant from communities or political queries. If prescribed fire is to be more than pile burning, it needs room—land to spread, land to absorb escapes and errors, space to dim close scrutiny. Otherwise, that nominal wildland will become a greenbelt, a kind of urban parkland, meticulously tended not to promote fire but to prevent it. Wildfire will be something to be managed like road reconstruction to protect the key assets, the lake's waters and its communities.

Where it lacks space, fire management will have to trade money for land. The 10-year strategy proposed for Tahoe came with an anticipated cost of $206 million to $244 million dollars over a decade. At a place known globally, and valued regionally and nationally, at a time flush with funny money as the mid-Noughts were, that was possible. Elsewhere it isn't. Most places in need of fire restoration don't warrant presidential visits or claim the attention of the leader of the Senate. Tahoe can. That's what makes it special and why it is not simply a cameo of the West but an apotheosis. The basin is a catchment for western history and fire; and what it gathered, it has arranged in unique and striking ways. But what came in is not likely to leave.

WORKING FIRE

THE RANGE OF LIGHT tilts downward from south to north and from east to west. It is highest at Mount Whitney in Sequoia National Park and lowest around Mount Lassen, where the Sierras grade into the Cascades. That is how the land lies geographically.

But the Big Tilt also defines a grade between wild and working landscapes and between comparable programs of fire management. To the far south a quarter of the Sequoia National Forest is lodged in legal wilderness and another quarter designated as Giant Sequoia National Monument. To the far north the identically sized Plumas National Forest has one designated wilderness at 2 percent of its holdings. The High Sierra boasts a program of aggressive nonintervention, sweating hard to let fires burn as freely as possible. The northern Sierra runs an equally aggressive program of suppression, still in the practical throes of the 10 a.m. policy. Between them they define the poles apart of Sierra fire.[1]

———

The earliest Americans to the Plumas region found a landscape rich in gold, timber, grass, water, and fire, and they worked them hard.

Early placer miners panned out the easy nuggets, then constructed flumes to channel the flow demanded by industrial hydraulicking, which gullied whole hillsides and sent a debris wave downstream to clog up the Sacramento drainage. Herders swarmed over the basins and slopes until their flocks ate out the understory. Loggers cut over what the miners left; only the Stanislaus rivaled the Plumas among California forests for the size of its cut; until, once again, the practice escalated into excess. After decades of high-grading, loggers began clear-cutting. Meanwhile, engineers moved from tinkering with stream flow for sluicing and hydraulicking into dams for hydropower, and finally to the massive damming and reallocation of rivers for the California Water Plan; the Plumas watershed furnished over 60 percent of the total. And fire? It was already lavish when the forty-niners arrived. Lightning frequently kindled dense busts of 30 to 60 fires from a single storm; the indigenous Maudi people had burned routinely and meticulously; and then the swarming Americans burned insouciantly and promiscuously. The Plumas became notorious as a fire-saturated forest.

The Plumas, in brief, was not merely a working landscape but too often a worked out one. Each excess brought a reaction, typically in the form of a denunciation by the outside culture along with imposed regulation. The "outside culture" was whatever competed for the economy of the day. The hydraulic mining controversy inspired an 1884 court injunction to shut it down and then congressional intervention in the form of the California Debris Commission, appointed in 1893, to regulate the practice—this at the insistence of Sacramento Valley farmers alarmed at watching a glacier of cobbles, punctuated by floods, crunch over their wheat fields and fruit orchards, and by commercial interests who feared for the region's rivers and for the integrity of San Francisco Bay. Preventing destruction of navigable streams was the constitutional authority that allowed for such "enormous interference" by the federal government in what would otherwise be a state affair. The light-burning controversy of the early 20th century put the federal agencies on the path of fire exclusion—this at the insistence of professional foresters who feared and despised the laissez-faire fire economy of the frontier. The spotted owl controversy in the early 1990s, leveraged by the Endangered Species Act, brought wholesale logging to an abrupt halt—this at the insistence of a service and amenities economy.

The Plumas has been at or near the center of it all. It has a lot to answer for.

Whatever the end commodity, the means to extract it on the Plumas usually involved water or fire. That a natural process would get redirected to serve human purposes is a norm, not news. What elevates the Plumas into a paradigm is that it is so richly endowed with water and fire and that it has continually injected them with what, to outsiders, or competitors, might appear as performance enhancers.

Water was the agency that collected nuggets into placers, buried them in Mesozoic times, and then eroded them back into play when the Sierra Nevada rose upward. Washing and sluicing were the controlled means to extract them in usable forms. More water meant more gravel washed, which meant more nuggets. A perverse economy of scale eventually captured and redirected into the flumes whole streams, and then dammed rivers. Small-scale operators could not compete, and the big money confirmed their monopoly by buying up the water rights. Debris dams tried to hold back the flood but quickly washed out. The amount of earth moved was estimated at six times the volume excavated in digging the Panama Canal.[2]

What finally shut hydraulicking down was the threat downstream. Towns like Yuba City had to raise levees, and then flooded anyway. The Plumas County seat, Quincy, had to be relocated when a debris dam broke and inundated the old town plat. Whole valleys filled with ancient cobbles to make glaciers of debris—a macabre double to John Muir's radiant ice glaciers. Across the greater drainage a debris wave sloughed like some rough beast toward the Sacramento delta. Ultimately, it threatened to seal off San Francisco Bay not by direct filling but by reducing the outgoing tidal prism of water, which would be so weakened that it would no longer be able to scour across the tidal bar outside the Golden Gate. An alliance of San Francisco commerce, Sacramento Valley farming, and the embryonic conservation movement succeeded in effectively closing the spigot.

The California Debris Commission did not end the story. Some illegal operations continued, and in 1905 the price of gold had so risen that the mining interests petitioned President Roosevelt to lift the ban as a pointless hardship. Roosevelt turned the task over to the U.S. Geological Survey for a full-bore inquiry. Science would rush in where politics feared to tread. By the time the report was released in 1917, it no longer mattered that its author, G. K. Gilbert, formerly the Survey's first chief geologist, revealed the full horror of hydraulicking. The price of gold had fallen; the debris wave through the Sacramento Valley had passed its worst phase; and the issue was moot. The future would continue to pick up after the mess. Some 140 years later debris is still sequestered in nooks along the Feather and Yuba Rivers, like enormous lithic stumps from a bygone era.

Still, the water flowed. Now, instead of directing it into flumes for mining, it was shot through generators. The old cascade of debris dams was replaced by one of hydropower dams. Then, as part of the California Water Plan, another crop of dams spanned the rivers. The electricity went to the new urban economy and the water, once again, to valley agriculture and on to the sprawling cities in the south. The rivers of the Plumas were working waters.

The fire scene underwent a parallel evolution. There was lightning fire aplenty, just as there were abundant rivers on the land. The indigenous peoples, the Maidu, domesticated part of that fire and used it to shape their hunting, foraging, and horticulture. They burned along corridors of travel to make passage easy and, in the words of a Mewok elder, "to prevent bear from hiding along such routes and to prevent enemies from ambushing them." They burned to promote shrubs useful for basketry. Where they collected seeds seasonally, they burned "around the oaks and nut pines to make acorns and nuts easier to gather." Surely they burned, as almost all such economies do, to encourage browse for favored game, and, undoubtedly, some fires bolted free. They burned early and burned light. Lines of fire, fields of fire—together they laid down a matrix of fuels that helped define what lightning fires might do. Patch by patch, tribe by tribe, the pieces added up to landscapes brought into a rude order of fire by the jostling push and shove of torch and bolt.[3]

Such practices kept wildfire within bounds, much as hunting did wild life. Then, as one forty-niner observed, "the whole world rushed in." The soil, water, forest, fauna, and indigenes were overturned with a rapidity that had few historical rivals anywhere. Within decades the indigenous fire regime was extinguished, along with the natives. The old regime yielded to a fire rush. In 1902 John Leiberg, surveying forest conditions in the northern Sierra Nevada for the U.S. Geological Survey, confirmed that the only constant was the unremitting presence of fire. "The most potent factor in shaping the forest of the region has been, and still is, fire."

It had preceded the white man, but the newcomers had drastically altered the terms of its presence. So outrageous was the promiscuous burning that Leiberg thought lightning was only a probable cause. Its presence, like that of Maidu fire, was overwhelmed by the onslaught. In effect, the fire regime had been ruthlessly overrun and seized, just as land and woods had. What followed was a kind of fire hydraulicking.[4]

Over the past few years, surveying one forest reserve after another, Leiberg had witnessed a lot of burned land throughout the West. He understood the dynamics behind the flames. "When the miners came, fire followed them." They left large tracts of chaparral along with debris dumps. Soon after came the "flock masters" and sheep. They were widely accused of setting fires "in all directions"; and all the fires Leiberg saw "closely followed the sheep camps." But everyone joined in: there was no order, the social upheaval was as profound as the environmental. The fires sprawled as much as the settlers over unpatented lands.[5]

A 1904 survey of the Plumas forest reserve summarized the matter-of-fact panorama of practices. "The white man"—picking up the torch dropped by the Indians, who were "accustomed to burning the forest"—has "come to think the fire is a part of the forest, and a beneficial part at that. All classes share in this view, and all set fires, sheepmen and cattlemen on the open range, miners, lumbermen, ranchmen, sportsmen, and campers. Only when other property [than timber] is likely to be endangered does the resident of or the visitor to the mountains become careful about fires, and seldom then." Much as with mining, Leiberg identified the worst wreckage as concentrated in a "fairly well-defined belt ranging from 15 to 20 miles in width and stretching diagonally across the entire region from the northwest to southeast." But everywhere he chronicled the melancholy legacy left by the flush times. The ruined timber: a third of the Middle Fork of the Feather River badly burned, and only 20,000 acres out of 586,450 acres not scarred. The loss of young growth: twice or more the present volume would exist save for the "enormous" destruction of reproduction over the past half century. The fire-scarred big timber that survived: each successive burn gnawed further at the bole until it felled them. The fire-scarred landscapes: the severe burns came back to brush, and each subsequent burn only "fastens the chaparral more irremoveably in its place."[6]

Yet there was a counterrecord that emerged not from academically trained elites but from folk memory. It noted that the fires were most damaging in mixed conifers and less so in open pineries. It recorded that, as along the Plumas, "the common type, and in fact almost the only type, of fire in the forest is a light surface burn that at first sight seems to do little damage." Such fires had become the norm—might well have been the norm before the rush. Because of their "frequent consumption of forest debris," large placers of fuel, "sufficient to furnish food for fierce fires, are seldom met with." Such light burns were viewed by locals as beneficial, and patches of woods that escaped them were fired to put them under that regimen. (Officials often found it difficult to prosecute for arson because, even when caught torch in hand, local justices of the peace and juries refused to convict due to a "peculiar attitude" toward fire.) The "forest" of the northern Sierra, that is, was indeed mixed and displayed a mixed stew of fire regimes.[7]

Still, the outcome seemed an extravagance of burning. Those committed to protecting the forests were little inclined to discriminate between careful burns and reckless ones, any more than the California Debris Commission tolerated much hydraulicking. Regulation was possible but difficult to enforce, and any administrative aperture merely invited practitioners to slip the leash of oversight altogether. Critics simply wanted fire controlled. Revealingly, the same time that the California Miners' Association petitioned to renew hydraulic mining and free up the potential bullion still buried in ancient soils, a debate arose over fire practices and their regulation. What followed became known as the light-burning controversy.

On one side of that intellectual fireline stood the local caucus of forest users, all of whom wanted access to fire as much as miners did access to water, and all of whom insisted that continued "light burning"—what they insouciantly called "the Indian way"—was the best means to protect existing timber. The big timber owners such as T. B. Walker and the Southern Pacific Railroad, with major holdings that graded into the Cascades, ardently supported them. Cease that practice and forests would become overgrown, susceptible to savage burns, insect attacks, and disease. They viewed formal fire protection, both fire prevention and fire fighting, as foolish, academic, and self-defeating.

Against that frontier rabble stood foresters, university science, and urban-based conservationists, among them members of the embryonic Sierra Club. While they admitted that some fires might be harmless, they sought fire's abolition. Their reasons were many. They pointed to forestry, which condemned all burning. They noted that seemingly benign surface fires destroyed humus, the biological capital of the soil, and burned into the boles of old growth, eroding away timber values. They cited severe burns which surrendered forest to brush, and then, by repeated burns, however light, perpetuated that type conversion. They felt a pluralistic message was too complicated for public consumption: a simple ban was easier to administer. (This was the same era that promoted Prohibition.) And they saw in the sneers and the unholy alloy of populistic protest and big money a challenge to public land and the very idea of a public commonwealth. This was national politics—a fireline drawn in Sierra sand.

The controversy became a political firefight in the summer of 1910. The Big Blowup traumatized the U.S. Forest Service, which inherited the forest reserves in 1905, while sparking public debate over policy. Northern California was the hotbed of protest. The California state engineer, W. H. Hall, voiced objections to the paramilitary suppression model. *Sunset* magazine, a subsidiary of the Southern Pacific Railroad, ran an article that turned the argument around and recommended using troops to light fires rather than fight them. Secretary of the Interior Richard Ballinger expressed, in San Francisco, his opinion that light-burning was the superior strategy. The battle was joined. The Feather River experimental forest became the first dedicated to fire research; it was where S. B. Show, who later ran California fire the way the Southern Pacific did politics, earned his spurs. The controversy ended formally in 1923 when a panel commissioned by the Board of Forestry reported on experiments in northern California that sided with the abolitionists. Light-burning became anathema, and light-burners were afterwards treated with the aloof scorn reserved for perpetual-motion mechanics.[8]

Excess had again led to regulation from the outside. The U.S. Forest Service assumed the role for fire that the California Debris Commission had for mining. Fire would not be regulated so much as suppressed. If Northern California was the insurgents' stronghold, so it would become the showcase for the Progressive alternative, systematic fire protection. Curiously, the "Indian way" of fire management passed the same year a Yahi Indian, Ishi, the last of his tribe, wandered through the watersheds of the Plumas to Oroville and surrendered to the authorities.

On the Plumas much of the forest was fire hardened and the worst incineration was localized to old mining and shepherding camps. Profuse pineries still rippled across the landscape. By the 1920s hydraulicking had yielded to hydropower and grazing to logging. The catalyst was the Western Pacific Railroad which punched through the Feather River Canyon between 1905 and 1909.

That main line grew tentacles into valley after valley. Steam and steel began working the forest as sluice boxes had streams. Through World War II the Plumas trailed only the Lassen National Forest for timber production in California. Big fires kindled in the initial slashings; but gradually better spark arrestors, more attentive maintenance by the railroads, and more organized firefighting knocked down the flames. During the Depression the CCC was particularly active on the Plumas. Quickly, the infrastructure for fire control fleshed out. Lop-and-scatter replaced slash pile burning.

Then came the postwar housing boom—more explosive in California than anywhere else in the country. The private timber holdings could not meet demand, and the Plumas cut more. Roads replaced rails. Influenced by a Cascades (Oregon) model, the forest shifted from selective cutting to clear-cuts. That left lakes of slash that had needed firing—what passed in the Plumas for prescribed burning. A 1988 forest plan anticipated continued high timber production. Logging now slid into a familiar regional scenario. Bigger cuts became more abusive; they clashed with other features of the landscape of interest to outsiders, particularly old-growth woods and endangered species. The spotted owl took the place previously occupied by navigable rivers and light-burning; the timber industry found itself slashed rather than doing the slashing. Lawsuits shut down six of the seven mills on the Plumas. The local economy descended into shambles. Violence simmered and occasionally broke out in what became known as the "timber wars." In response, the Forest Service labored to produce an environmental impact statement to address issues over the Sierra as a whole. In 2001 the Plumas was absorbed into the resulting Sierra Nevada Forest Framework.

———

Once again the outside world had intervened to halt what it regarded as wreckage, and true to form, the Plumas sought to claw back some space to keep it a working landscape. The upshot was the Quincy Library

Group, first organized by an industry forester, a county supervisor, and an environmental lawyer in late 1992, which sought to thread the labyrinth between timber harvesting, the California spotted owl, and local communities dying on the vine. The pivot was fire protection.

The Quincy Library Group grew into a community caucus and then went national, getting some traction with the Forest Service and funds for pilot projects. But the old firefighting adage—force enough, fast enough—seemed to get reversed. The scale, both in time and space, was too small; wildfires would take what axes couldn't. The group appealed to Congressman Wally Herger and Senator Diane Feinstein, which led to the 1998 Herger-Feinstein Quincy Library Group Forest Recovery Act. Going national and political, however, made the opposition do likewise, as the timber wars went to the pages of the *New York Times*. Still, the act passed, with funding for five years. Its model contributed to the National Fire Plan (2000) and Healthy Forests Restoration Act (2003), with their emphasis on fuels management as a core project. The act has been extended twice.

What brought antagonists together was, as one observer termed it, a "shared sense of desperation." The county faced poverty, and its forests stared at the prospect of larger, more severe fires with a downsized Forest Service and without the old woods industry to supplement fire control forces. In principle those interested in preserving habitat and those who wanted to log might find common cause in a shared alarm over a future of conflagrations that served neither. The solution was to identify strategic locations for fuelbreaks, what were termed "defensible fuel profile zones"; to thin the overgrown understory and convert it to ethanol; and to restore the bad burns by means that included timber salvage. There were special protections hardwired into the agreement to shield riparian zones, California spotted owl sites, and old growth generally, and to rehabilitate watersheds. All this could revive the local economy, not least its logging industry, though on a much smaller and nimbler scale.[9]

The Plumas of course emphasized extraction and commerce. But the sense grew throughout much of the northern Sierra Nevada that fire was something on which rivals otherwise locked in death-spiral vendettas might find a collective purpose. Whether or not they logged, fished, hunted, hiked, erected shacks or log-plated mansions, wanted to wipe out endangered owls or throw themselves across the path of feller-bunchers, everyone who lived in the region had to cope with fire. When he was a young man, working as a fire lookout and occasionally fighting fires, poet

Gary Snyder liked to brag that he had found a higher calling. Then came the disillusionment as he viewed the great firefight as a Cold War avatar and denounced the Forest Service for its mishandling of the Sacred Cub (Smokey). But when he moved to Grass Valley, he realized that bad fire was a threat to all. It was something around which people might convene, as they would a campfire.[10]

That was not the sentiment of the Plumas. The big fires continued. The Moonlight fire (2007) started from logging on private lands. The Storrie (2008) and Rich fires began from railroad maintenance crews. The Canyon complex (2008) absorbed 12 lightning strikes into one converging vortex. The hoary pattern of fires along corridors and in patches of worked land persisted, although the matrix was no longer dense enough to contain the lightning busts for which the northern Sierras are famous. The consensus is that the fires are getting worse and that however useful the defensible fuel profile zones, they are inadequate to the task. They are too few for landscape-leveraged conflagrations; their construction is much slower than fire's spread. The Plumas maintains a full suppression policy.

So the working fire endures, though not so much as a means to extract resources than as a resource itself. Fire's management has become a means to keep the local economy ambulatory. The region has struggled to make the transition from logging camps to retirement condos, from commodity production to amenity services. In 2011 Quincy had an unemployment level of 23 percent. The economy of the forest, like that of its towns, depends on earmarks and public subsidies.

One reading, the world-weary, is that the Plumas remains whipsawed between excess and repression. Another, more hopeful, is that the region will, somehow, overcome not only corporate greed, bureaucratic inertia, and environmentalist outrage but its own history and rediscover the American pragmatism that was also formalized at the time of the light-burning controversy. If so, if some means can be found to keep the land working without being worked over, almost certainly the Plumas will remain at one pole of Sierra management; but it will be a pole that can share some charged field with its opposite. Fire will likely be the medium of exchange.

THE PASTURES OF PURGATORY

T HE CALIFORNIA INDIANS had no livestock. Instead they hunted wild fauna—deer, rabbits, bushy-tailed woodrats, even grasshoppers; and they used fire extensively. Sometimes they burned for fire drives, as Captain Fernando Rivera y Moncada described at Monterey in 1774. "The heathens are wont to cause these fires because they have the bad habit, once having harvested their seeds, and not having any other animals to look after except their stomachs, they set fire to the brush so that new weeds may grow to produce more seeds, also to catch the rabbits that get confused and overcome by the smoke." Often the natives burned for habitat, and their preferred species shared in the refreshing of browse and grass that accompanied firing for acorn harvests, shrub regeneration, and general land cleaning.

Their fire practices resembled those of Australia's Aborigines, famous for "firestick farming," a style of fire-catalyzed living so intensive that it mimicked a kind of landscape horticulture. They burned shrubs to promote basketry twigs, they burned covert to keep it low and scattered, they burned oak savannas to disinfect acorns from larval filbert weevils and white moths. And like their Australian counterparts California's indigenes saw that world shatter not simply because Europeans invaded but because the colonizers brought livestock.

The orders that founded Mission San Juan Capistrano instructed the colonizing padres to bring nine milch cows, a breed bull, a yoke of oxen, eight pack mules, three saddle mules, three horses and two mares, two pigs (a boar and a sow), and whatever chickens San Gabriel mission could give. But almost immediately the needs of introduced stock clashed with those of the native fauna. Revealingly, the Spaniards under Captain Rivera fought that harassing fire to save pasture for their stock. The mission demanded that California Indians surrender their old way of life to the discipline of mission, farm, pasture, and ritual; so, too, wild animals would yield to domesticated ones; and inevitably aboriginal fire regimes would break against the seawall of a new order.[1]

Cattle, sheep, goats, donkeys, horses, swine—all extended the reach of Spaniards, Mexicans, and Americans far beyond their grasp. The newcomers' "portmanteau biota," as Alfred Crosby termed it, also loosed a witch's brew of allied species that collectively remade California ecosystems. Wild oats and mustard, long allies of European fauna, as quickly replaced the native annual grasses as cattle did grizzly bears. The grazing regimen went from the wild to the tamed or, given the looseness of herding, to the feral. Exotic grasses, overgrazing, and trampling, compounded by drought, all connived to do to pasture what hydraulicking did to rivers and logging to forests. Another surge followed the gold rush. By 1862 there were an estimated three million head of cattle and maybe half a million sheep, all rapidly reproducing, throughout the state. The free-ranging herds did to native fauna what forty-niners did to native peoples.

The newcomers' livestock often thrived better than the newcomers themselves. With the secularization of the missions in 1833, large land grants, organized into rancherias, replaced missions. The hide-and-tallow industry became the mainstay of maritime commerce beyond Alta California. Cattle spread throughout the Coast Ranges and into the San Joaquin and Sacramento Valleys. As the range degenerated, or brush crowded out forage, ranchers seized the torch from the faltering indigenes and put it to their own purposes. There was less burning than before, and the fast combustion of fire now had to compete with the slow combustion of rangy herds, but fire persisted. When the interior valley flooded in 1862, followed by two years of drought, herders turned to sheep and moved into the Sierra. Wherever they went they burned.

And burned again. Their trailing smokes signaled the arrival of a new matrix in California's fire regimes. Cattlemen burned to keep down the brush or convert brush to grass. Shepherds trailed a spoor of fire as they migrated down the mountains each fall, transplanting the *trashumancia* of Mediterranean Europe to the American cordillera. Surveying the San Jacinto Forest Reserve, H. R. Porter Jr. wrote, "The whole area is covered with chaparral and subject to overgrazing and fires. The signs of fire having gone through the brush are constantly evident, and smoke can usually be seen. Cattle, sheep and goat grazing is carried on to the limit of the range and beyond. The men riding the range freely acknowledge that burning for the purpose of improving the range has been carried to such an extent as to have very decidedly injured it." In the Sierra Nevada John Muir thundered against sheepherders' fires, as though the loathed "hooved locusts" could strike sparks when they flocked: "Running fires are set everywhere, with a view to clearing the ground of prostrate trunks, to facilitate the movements of the flocks and improve the pastures. The entire forest belt is thus swept and devastated from one extremity of the range to the other. . . . Indians burn off the underbrush in certain localities to facilitate deer-hunting, mountaineers and lumbermen carelessly allow their campfires to run; but the fires of the sheepmen, or *muttoneers*, form more than ninety per cent. of all destructive fires that range the Sierra forests." W. J. Lord penned a more considered analysis from Tuolomne County: "Burning at that time became such a practice that people knew when sheep were leaving the mountains by the number of fires set. Smoke . . . was so thick at times it was hard to see at midday. No attempt was made to stop the fires unless someone's place was threatened, then back fires were set and usually the fire went some other direction. These fires burned thousands of acres almost everywhere where timber and brush grew in the mountains." Yet most fires burned benignly on the surface.[2]

All was not pandemonium and havoc. Pastoralism changed the land, which in turn changed fires, which further altered the character of pastoralism. Where land was fenced and husbandry the norm, fires were constrained; they resembled the spring burning that swept away the strewn branches of well-pruned orchards. But where the estates were vast or the flocks ran over the public domain, the fires could free-range, subject to occasional roundups and brandings. It took much of the 20th century to fence in those herds and their enabling fires. The creation of national parks and forest reserves throttled the most abusive pastoralism in the mountains, as they did migratory logging. By the 1920s ranching was receding to private lands, although these could be vast, either through the legacy of land grants to Spanish grandees or railroad magnates, or by very cheap purchases, the manipulation of land laws (by declaring sites as "swamp"), or outright fraud. The big ranchers, along with timber owners and corporate interests (including the Octopus itself, the Southern Pacific Railroad), were the landed gentry of California. They supplied the beeves for what V. L. Parrington called the Great Barbecue, the lavish liquidation of the public domain into private hands for a pittance. By the end of the century 90 percent of grazing had migrated onto private land. The most productive ranching was where agriculture also best flourished, the Central Valley. The rest ranged along the foothills of the Sierra Nevada and amid the inherited rancherias of the Coast Range.

Little of this scenario was unique to California. What California did to ranching was what it did to everything else: it bulked it up. Much as with light-burning, this was where the conflict to regulate free-ranging livestock over the public domain was exaggerated and then resolved. In fact, both issues, burning and herding, climaxed in almost same year. The light-burning controversy flared into public discourse in 1910. The test case for Forest Service regulation of grazing, originating in California, was settled by the Supreme Court in 1911. California was the political corral where fights over pastoralism burning in the West were fought out.

The Mendocino National Forest is an anomaly. North of San Francisco the Coast Range thickens like strands on a rope. The Mendocino lies on the eastern, drier side of that bulge, but its biota still more resembles the Pacific Northwest than the Coast Ranges below the bay, and forest management falls under the aegis of the Northwest Forest Plan. Its eastern flank, however, grading into the Sacramento Valley and subject to valley thermals, is encrusted with hard chaparral that, as the land sinks lower, gradually thins into oak savanna and eventually shakes off its woods altogether to become tumbling hills of grass. Those two biomes, wet forest and dry brushland, meet at the eastern ridgeline. The range even has a unique sundowner wind, but revealingly it is reversed, flowing from the coast inland. The Mendocino is the only national forest in California without a paved road across it.[3]

The lower landscape is a natural for grazing, and ranchers have sought to expand that domain by pushing back against the mountain brush.

On the summits were glades that they routinely burned to keep fresh with browse, and from which they let fires spread. Like herders everywhere they resisted any effort to rein in their unrestricted access; and in the mountains, physical access was only as good as ecological access, which for herds meant burning. Unsurprisingly, the Mendocino, and the lands south of it in private ownership, were loud in their clamor for range burning, merged their case with the light-burning arguments from timber owners to the north, and when that failed, became notorious for incendiarism. But brush burners didn't need to rely on surreptitious slow matches. Fires routinely spilled over from private lands with calculated indifference. The important thing was that the chaparral-clogged slopes got burned. To the Mendocino the Forest Service assigned its toughest fire officers.

After World War II the pressures for burning intensified and found an institutional point of compromise. Ranchers organized into a Range Land Utilization Committee and demanded more range improvement, for which a match was the cheapest tool in the shed; other groups signed on in the hopes of improving watersheds or wildlife habitat; and they found an academic voice when Harold Biswell joined the faculty at the University of California, Berkeley, and, fresh from experiences in the Southeast, ardently promoted prescribed fire (which at first, in a colossal gaffe, he called "light burning"). Professional rangeland managers agreed that "brush conversion" was "about the only way to increase materially the present area of rangeland." The livestock industry saw conversion as "an opportunity and a challenge" of the first magnitude.[4]

The fulcrum to leverage enthusiasm into practice was a law enacted by the legislature in 1945 that authorized the Division of Forestry to issue permits for burning on private lands, to enter into contracts or cooperative agreements with individuals and cooperatives to burn on lands for which the state had primary responsibility for fire protection, and to sponsor research and experiments. On behalf of the Board of Forestry H. L. Shantz summarized existing knowledge, argued for site-specific practices that distinguished between the "brush" in redwood regrowth and the "brush" of chamise-dominated chaparral, and fretted over those "who look upon fire as a cure for all their troubles," yet confirmed the consensus that "the encroachment of chaparral upon both agricultural and grazing lands" was an issue of "immense importance" and could not be resolved until fire was properly managed. There was no argument about the value of fire when used correctly, only when and where to apply it.

In 1951 the Forest Service Regional Forester, Clare Hendee, articulated that agency's policy: its first charge was to protect watershed and soil, but burning could be used in careful coordination with CDF and the Range Land Utilization Committee or local Controlled Burning Committees of landowners. The USFS would not, however, burn on private lands, and private landowners who sent fires into national forests would be held liable. Between 1945 and 1952 some 700,000 acres were burned, although 91,000 acres came from escapes.[5]

As that first wave ebbed, Biswell sought to carry fire from the scrub to the timber by creating a demonstration site at Hoberg's Resort in Lake County, immediately south of the Mendocino. His message was received by students and later by the Tall Timbers fire ecology conferences, but got scant traction elsewhere; what mattered for range burners was the brush. That scientific message was summarized in 1954 by Arthur W. Sampson and L. T. Burcham's *Costs and Returns of Controlled Brush Burning for Range Improvement in Northern California*. They confirmed that the practice worked—that was the good news. The bad news was that only a fraction of chaparral landscapes were amenable to genuine conversion, burns yielded mixed results (particularly if not seeded afterward), and each burn demanded planning, staffing, equipment, and supervision. As with light-burning before, the economics, they concluded, argued against expansive practice.

Meanwhile, fires escaped. Tort law did what fire prevention hectoring could not; enthusiasm cooled. In 1957 the public agencies came to a similar conclusion when the Klamath National Forest, after careful preparations, kindled 500 acres of smashed brush, and watched the flames scamper over nearly 13,000. That misstep quenched Forest Service ardor for burning for the next two decades.[6]

The exception was the Mendocino. It had a history of experimental tinkering, partly in response to local pressure and partly from the forced prompting of large wildfires; and it could count on local support since its chaparral rubbed against a stubborn local culture of burning. From the origins of organized fire protection in California it had served to test, demonstrate, and propagate. In the early years it was where ranchers and foresters squared off over deliberate burning. In the postwar era it was where they found, along with hydrologists, wildlife biologists, and surprisingly fire control officers, common ground in well-burned land. Suitably fired brush would improve forage, increase useful runoff, and assist fire suppression.

The 1947 Schuyler fire served as a catalyst. Soon afterwards the forest enlarged its ambitions and entered into a cooperative agreement with the state Fish and Game Department to administer 12,000 acres for deer habitat. A second jolt came in 1953 when the Rattlesnake fire burned over a crew of 15 amid the chaparral-encrusted ravines that all parties wanted to thin. The upshot was that the Mendocino became the scene of active research, a counterpoise to Biswell's plots at Hoberg's. This took the form of the Grindstone Project, an expansive program of chaparral "management," which translated into patches of type conversion and fuelbreaks along ridges and soon became a demonstration site. In Robert Cermak's words, the Grindstone Project "led the way in chaparral management for the entire state." Samuel Dana and Myron Krueger, respectively emeritus dean and professor at the University of California, Berkeley, declared that there could be "no question as to [the] urgency" of the brushland issue throughout the state.[7]

Yet even as it hacked, sprayed, and burned over the ridges leading to Doe Peak, applications for permits subsided and, more precipitously, applications for reburns fell. Historic rangeland burning dropped faster than the grade of Grindstone Creek. The reasons are many, but a big one was that not all fires worked and another that not all brush was equal and interchangeable. One "brush field" might be improved with fire, another worsened. Because of past fire history brush flourished in some areas that might support trees or grass; and for such sites keeping fire out or shrewdly putting it in could restore the old regime. But for many areas chaparral was the natural and durable cover. Burning to convert it to something else did little good, and the attempt was costly. The price of escapes was potentially ruinous.

Still, while brush burning struggled across the state, it flourished at Grindstone because many communities wanted it. The fire community itself—so often skeptical if not hostile to landscape burning—sought a successful program of controlled fire. Wildlife habitat and fuel abatement could substitute for range improvement as justification; public lands were not subject to the same economic pressures as private; and the memory of the Rattlesnake fire still festered. From using fire to manage chaparral, interest shifted to managing chaparral to control wildfire. At Grindstone Canyon it seemed possible to bond enduring local desires to the national trends of an emerging fire revolution.

An era of formal demonstration ensued. In 1973 the Forest Service, California Department of Game and Fish, Glenn County, and local ranchers joined forces to crush 1,500 acres of brush and burn 12,000 acres within a matrix of fuelbreaks along 100 miles of ridgetops. Mission creep carried the program up from the brush into the timber. Plans called for rotational burning every 10 years on grassy fuelbreaks and 20 years on resprouted chaparral. The fire-fuel cycle and the fire-forage cycle would come into sync. Subsequently, the California Division of Forestry and the Soil Conservation Service joined for yet another renewed cycle, this time in 1981. By now the Grindstone's role as a demonstration site included firing by helitorch. Enthusiasts noted that costs of treatment had plummeted from $85 per acre in 1973 to $3 to $5 in 1980. The announced goal was restoration to something like presettlement conditions. The Grindstone became the flagship of six demo sites in and around the Mendocino. It was hoped that the lessons learned could be applied elsewhere, not only in Southern California but "in other parts of the world where conditions are similar." In 1982, partly based on the Grindstone experience, California replaced its range burning law with a more wide-ranging chaparral management act.[8]

For most of the 20th century, in brief, Grindstone Canyon has persisted as a historical vortex. Because of the early rancor over folk burning, it became a hotspot in the saga of systematic fire protection. Because of the Rattlesnake fire, it acquired some of the cachet attached to other tragedy sites like Blackwater Canyon and Mann Gulch. Because of the ambition, in the formative years of the fire revolution, to reinstate fire and restore California to something of its presettlement status, it was proclaimed an exemplar for demonstrating how fire ecology and fire behavior—ecological burning and fire control—might converge. Times changed as fire control morphed into fire management, but Grindstone was where new generations of fire officers could break a lance or two as they sought to solve the fire and brush conundrum, like mathematicians drawn to Fermat's Last Theorem.

In an odd way the scene behaved as though administrative history was recapitulating fire history. Demonstration projects returned with the seeming frequency of slope-sweeping chaparral fires. Every couple of decades Grindstone reincarnated into an avatar better suited to its times. In this transformation there was little mystical. Because it didn't stand cheek by jowl with a metropolis, it could tolerate experimentation, field trials that neighbors cheered on rather than shut down. Because of past experiments there existed a body of research data and, no less critically, an infrastructure, that made it relatively easy to start over, to clean out the old tracks and direct them to new purposes. Pilot programs for fire management resprouted with the vigor and regularity of chamise.

But then reinvention is an old California trope.

At the onset of the postwar era it seemed that California might emulate its bicoastal counterpart, Florida, where range burning held fire on the land. The 1945 Rangeland Improvement Act encouraged controlled burning. In 1958, four years after the peak of the postwar boom in permits authorized under it, Harold Biswell published an extended argument that what was good for Georgia was also good for California. The two groups that would most likely keep fire on the ground, ranchers and foresters, had signed a cease-fire. In overgrown brush they had found common cause.[9]

It didn't happen. As always the reasons are several. California lacked a cultural taproot for burning. Even in the postwar era those who owned land in Florida assumed you burned unless something compelling stopped you. Most burners were surrounded by others of like mind. In California the big landowners assumed you didn't burn and had neighbors who thought the same. Its culture of burning, like its grasslands, was not based on perennials but annuals. The deep-rooted Florida fire culture could weather the bad seasons. California's needed constant refreshment.

Then there were ecological considerations. California's indigenous grasslands had been early savaged by European colonization; Florida's interior remained largely untouched. In Florida unburned land quickly regrew, thickened, and became impenetrable; five unburned years might render it unusable, and a decade, the scene for explosive fires. Florida ranchers needed to burn annually to freshen pasture and hold the ornery rough at bay, and they found ways to keep burning until they passed the torch to public agencies. In California unburned land remained as seemingly quiescent as the weather. Degradation resembled an ecological wasting disease rather than a lethal fever. Ranchers could skip a year or two, or even a decade, of burning. While they fretted over thickening brush, they continually found ways to keep from burning. Increasingly their neighbors wanted fire banned or demanded so many restrictions that firing became hopelessly cumbersome and costly.

In Florida the public and private economies of fire converged. After some false steps the Florida Forest Service buttressed the argument that burning was a landowner's right. It could even intervene on private lands and burn where the vegetation presented a hazard. In California, whatever their intentions, public and private diverged, and when joined, could not resist the outside pressures. CDF could send honor camp crews to do work that ranchers could never afford. The Forest Service could command national resources, and suffer the occasional breakdown, that local landowners could not. In 1982 California sought to transfer some of the strength of the public program to private lands. It replaced the 1945 Rangeland Improvement Act with a more expansive chaparral manage-ment program that authorized CDF to contract with private and public landowners on state responsibility lands to do the burning. CDF would share costs and its forces would put dozers and torches on the ground. If an escape occurred, it would assume liability. The private sector, where most ranching resided, could tap options previously available only to the public sector. The old constraints—costs, capability, tort claims—were cleared away. But law, markets, and public opinion still turned against range burning. The numbers continued their downward slide. In the early years annual range burning amounted to 140,000 acres per year. By 2000 "vegetation" burning overall was 20,000 and falling.

The fact was, small ranches could not compete economically against the big ones or challenge imported beef and mutton. They could not swap herbicides for burns. Air quality in the Central Valley was often obscene, and while smoke was not the worst offender, it was the most visible. Predators took more calves and lambs. Public lands were hostile to fires that slopped over borders. Suburbanites and exurbanites, if they tolerated fire, wanted it far from their fence lines. As ownership of surrounding lands slipped away from ranchers, so did their grasp on the torch. And that is the core reformation: California increasingly segregated between the public wild and the private city. Its fires reflected that split—restored on the public land, banned from the private. In Florida ranchers held that vast rural middle; in California, the middle suburbanized. The old fire border inverted. More and more, fires from public lands threatened private ones outside. The carrots for good burning got smaller, and the sticks for bad burns got bigger.

Still, range burning did not go quietly into the night. CalFire's new vegetation management program allowed it to conduct burns on state responsibility lands, not just issue permits, which also left it accountable for fires that slipped their leash. A California Prescribed Fire Council expanded the population of fire partisans, and fire associations could create a shared pool of labor and machinery. But it was too little, too late. The new landowners, descendants of the pioneering patriarchs, rode BMWs and raised houses. Transhumance became mechanized; the new "hooved locusts" of the backcountry were off-road vehicles. The corded Coast Ranges resembled a boa constrictor, bent on squeezing pastoral fire. Every gasp for breadth by ranchers or change of ownership or new escape or additional regulation allowed the economic boa to tighten its coil.

By the late 1980s some 90 percent of livestock was raised on private land, mostly in the Central Valley. Instead of moving cattle to the mountains in the dry season, modern ranchers shipped them by rail or truck to the Northwest; they fattened the herds on feed lots. Elsewhere, private ranches were not converting chaparral to grass but transmuting rural landscapes into urban. The threats came less from predation by feral dogs than by feral developers. In Florida development rimmed the state's beaches and only later dappled the interior, allowing controlled burning some elbow room. In California postwar development flourished amid the sites that were the prime free-range landscapes for herding. In the Central Valley ranchers sold out to irrigation agriculture. In the Coast Range they sold out to developers. Unlike Florida or the Flint Hills, California ranching did not become a holding company to keep fire on the land. Instead ranching withered; it sank under flood irrigation; it subdivided away. Successful ranches were either very large or became feedlots.

The former decided it could do without fire and the latter that fire had no place.

Pastoral burning found itself condemned, slight by slight, each act trivially venal by itself, to a purgatory of landscape fire.

═══════

In *The Pastures of Heaven* John Steinbeck tells the story of an emigrant who arrives from The East, anxious to create a dynasty, or at least to transfer a sense of landed gentry. In a Coast Range valley that an astonished Spanish soldier had named *los pasturas de cielo* (the Pastures of Heaven) Richard Whitehead erects a large house and plans a family to occupy it. "Husbandry" toward land was then an anomalous idea. "Few people in California in that day felt a responsibility toward their descendants." Neighbors warned that a house built for 500 years was "not how we work out here." You build a shack because, sooner rather than later, you "might want to move."[10]

Over the years everything turns against Whitehead. His wife has only one child, John. That son marries poorly and also has one son. The ranch gradually sinks into genteel disrepair. "In the West, where, if two generations of one family have lived in a house, it is an old house and a pioneer family, a kind of veneration mixed with contempt is felt for old houses." The American style is to build "flimsy houses and soon move on to some new promise." The son, Bill, has scant interest in either house or ranch. He and his wife move to town. Eventually, John decides to restore the land to productivity. "'Burn it off,'" says the hired hand. "'If you burn that brush this fall you'll get fine pasture next spring.'" The burn goes smartly until a small whirlwind scatters embers like chaff; somehow a spark lodges in or under the house; the house burns to the ground.

That, in cameo, is the story of California's ranching society. Most ranchers of course didn't burn themselves down. They couldn't burn—the threat of a calamitous escape was enough to dampen enthusiasm. But neither could they survive. A service economy did what fire suppression couldn't. Most of their descendants headed to the city and allowed the city to take over the ranch. They just sold out, as the Whiteside neighbors had always insisted they would.

THE BIG ONES

UNTIL MODERN TIMES city and country had always shared fires. American cities were typically rebuilt forests and burned with the panache and patterns of their surroundings. Countrysides were shaped by the demands of cities, both near and far, with slashed landscapes more prone to explosive fires. The same winds blew over both. The same logic of protection led each to prevent errant sparks, build firebreaks, and quench flames quickly.

During the frontier era, the two realms blurred. Then each matured and the fire scene calmed, as conflagrations disappeared from metropolises and were harnessed into the tamed rhythms of rural burning. More recently, in what might be termed a pyric postmodern phase, an outmigration of urban folk has begun recolonizing the countryside. This new frontier broke down the firewalls that had separated urban from wildland fire. The intermixing stirred by sprawl occurred everywhere, but it happened with particular vehemence in California, and most spectacularly, it burned. Urban fire returned, like an ancient plague once thought extinct and now revived in a more virulent mutation. City and countryside had to cope with fire along their shared fringe.

That was true north as well as south. The San Andreas Fault is less a geologic cut, neatly cleaving the crust into two sides, than a swarm of breaks as deep stresses release their strain and ripple through heterogenous rock. Even so, two regions along its complex trace stand out. To the south the swarm flexes into the Transverse Range. To the north a parallel fault swarm, highlighted by the massive Hayward fault, doubles the zone. The San Andreas proper runs to the west, and its echo, the Hayward, to the east. San Francisco Bay lies between them. The two dominant urban fires that frame California's fire century face each other with one of those faults at their back.

In the three centuries that span from Jamestown to San Francisco's immolation, a westering population had hastily erected towns and then watched them burn. The process of squeezing flame out of cityscapes came slowly and fitfully. Boston and New York continued to burn well into the 19th century. Only as buildings became less combustible (more brick and stone than wood), as flame became less abundant in daily life, and as firefighting became better organized did conflagrations reluctantly leave the metropolises and join footloose folk on the frontier. The last major urban fire on the East Coast was Baltimore's in 1904. San Francisco's fire two years later was the final conflagration of the settlement era.

Historians have come to think of the 20th century as a "short century," defined by the onset of the Great War in 1914 and the end of the Cold War in 1991. For fire history the corresponding events are the San Francisco conflagration of 1906 and the Oakland holocaust of 1991. The two cities pair off across the bay; so, too, they bookend a century. They are the Big Ones that haunt the imagination of California fire. They were big not because of geographic size but because they slammed into cultural centers that could register the shock. They killed. They incinerated, for the one, the major entrepôt of the western United States and, for the other, an elite community. They were fires that burst through nightmare into fact.[1]

In many ways San Francisco's fire history echoed that of the nation. The embryonic city was incinerated on Christmas Eve of 1849. The next year three fires swept through what was a high-order mining camp masquerading as a town. In 1851 the Great Fire destroyed three-fourths of the city, the same proportion as the 1906 fire. What the fire missed a second fire claimed later in 1851. Then, as the city passed through its accelerated adolescence, its fire scene settled down. Outbreaks more resembled episodes of domestic violence or saloon brawls than out-in-the-streets rioting. The big fires moved out to prospect the countryside.[2]

The mature city erected fewer combustible buildings and disciplined fires through a vigorous fire department. Paradoxically, without major burns to cleanse the cityscape—a kind of creative destruction—the city housed more and more relic structures from its wooden age. In October 1905 the National Board of Fire Underwriters investigated San Francisco and reported that 90 percent of the city's structures were wooden framed. No other major metropolis approached that figure, but then no other had progressed so suddenly from canvas tents to an edifice complex with monuments such as the Ferry Building and new city hall. Only 54 years before, the city had consisted of surveyor stakes in windswept grass.

When, at 5:12 a.m. on the morning of April 18, 1906, the San Andreas ruptured and sent earth waves rolling across the bay area, like a hurricane driving high seas before it, the city was vulnerable to fire; but it was not living on a knife-edge of conflagration. Rather than slow down settlement or compel rebuilding to national codes, the city had relied on its fire department to halt fires before they became large. For decades that strategy had worked because the number of fires was small and access relatively easy. An earthquake, however, changed that calculus. It kindled many ignitions, its rubble and ruptures checked movement, and it broke water mains. Firefighting alone might not succeed, and in fact, in the hands of the military and vigilantes who eventually took over the process, it proved fatally flawed as dynamited buildings and clumsy backburns almost certainly encouraged fire spread.

Given the density of the built landscape, it did not take much for stubbornly established fires to take out city blocks and then most of the city. A former chief geologist of the U.S. Geological Survey, G. K. Gilbert, was in Berkeley when the tremors struck. On Friday he took the ferry across the bay and recorded the fire's movement. "The westward progress of the fire north of Market Street has been checked chiefly by backfiring at Van Ness Avenue." Houses across the street were "blistered and had glass broken." The fire had broken across at one place before being checked at Franklin Street. "Backfiring is in progress N. of Pac. Avenue, and apparently being carried to the waterfront." From that point "the fire rushes up the slope of Russian Hill, consuming block after block of houses—chiefly of wood. The flames work with wonderful speed. While I lingered, whole squares were consumed. An hour is probably enough to raze a square of wooden houses." Ever the scientist—when the tremors had awakened him, while still in bed he began timing the shocks and analyzing their direction by the swing of the chandelier—he measured the burning time for a two-story house. "Roof gone in 7'; first falling of wall in 9'; flaming ruins in 12'."[3]

This was primarily a fuel-and-ignition fire. The April weather was warm and fog free, but California's parching summer had not started, the city was not gripped by drought, and the winds were mild. Synoptic systems moved through the region, slightly out of phase with the fires. There was an episodic easterly breeze, and the first day, Wednesday, the winds shifted confusingly between damp northwest and dry east flows. The second day saw the winds blow from the east, drying the scene and driving fire away from the bay. On Friday the northwesterlies returned. On Saturday it rained. There were terrain effects, notably as fires moved up San Francisco's fabled hills. But mostly what propelled the burn was the density of available combustibles and the abundance of ignitions—those that started the initial outbreaks and those set deliberately or ineptly in attempts to stop it. The critical winds were apparently those created by the fire itself.

The 1906 fire was a plume-dominated conflagration, as Jack London reported from the scene. "It was dead calm. Not a flicker of wind stirred. Yet from every side wind was pouring in upon the city. East, west, north, and south, strong winds were blowing upon the doomed city. The heated air rising made an enormous suck. Thus did the fire of itself build its own colossal chimney through the atmosphere. Day and night this dead calm continued, and yet, near to the flames, the wind was often half a gale, so mighty was the suck." When it ended, all that survived was "the fringe of dwelling-houses on the outskirts of what was once San Francisco."[4]

The defining event was neither fuel nor wind but a 7.9 earthquake that overwhelmed the fire protection system of the day. Too many fires, too little water, too much social chaos—that was the fire triangle that shook San Francisco to its foundations. For the rest of the 20th century urban fires in developed countries followed a similar scenario. It took earthquakes, wars, or riots to break down the built landscape and its social infrastructure sufficiently to where it could carry fire. As the century progressed, the San Francisco scene was turned inside out as urban fire moved to the fringe and left an incombustible core. Cities were subject to occasional catastrophes in high-density structures but no longer to free-burning conflagrations.

The final tally reckoned 4.7 square miles burned—508 city blocks, or 2,832 acres. Some 28,188 structures were lost (88 percent of them wooden). The official roster of fatalities was 700, but observers believe the likelier total ran between 2,000 and 3,000 lives. The official narrative deliberately turned away from the earthquake as cause, since it was a geologic precondition that residents could do nothing about, to the fire, which they could. For the next century the city that had begun in a de facto fire rush ceased to burn.

Refugees poured across the bay into Oakland. The saga of Oakland's fire history thus began as an aftershock of San Francisco's. The newcomers met a fire scene not unlike San Francisco's before it urbanized. The landscape consisted of grassy hillsides dappled with copses of trees, often redwoods, nestled near springs. The golden hillsides burned regularly and harmlessly. As long as the city lined the wharf or served as an outpost to San Francisco and Berkeley, its urban fires were those typical of wooden towns everywhere. What changed the dynamic was when houses moved up the hills.

The East Bay was a miniature Transverse Range. The Hayward fault shouldered up an embankment of hills. A seasonal wind, known as the Diablo since it flowed past Mount Diablo, a prominent monadnock, was a northern Santa Ana ready to spill over that height. Settlement evolved sentiments to preserve the framing hills and East Bay backcountry as natural preserve. All this replicated the basics that made the Southern California fire scene so explosive, and as the city crept from the bay to the hills, it promised to recreate the south's cataclysmic fires. What checked this scenario, however, was the absence of the high-octane chaparral that made blowing-and-going fires along the Transverse unstoppable. The Oakland Hills had annual grasses, savanna oaks, and riparian redwoods. Fires might be frequent and annoying; they would not be ruinous.

While San Francisco hastily rebuilt, in Oakland refugees became residents, and the city grew as an alternative. By 1910 it had ballooned to a population of 150,000. Developers—the new forty-niners of California prosperity—knew a "good sign" when they saw it. On the hills, arrayed like an amphitheater that looked out on the bay and San Francisco, they erected the Claremont Hotel, platted some 13,000 acres of land for suburbs, and began softening the windy grasslands with trees, shrubs, and ornamentals including broom that would add color, texture, and privacy, and (so they argued) might serve as fire windbreaks. They planted Monterey pine, a native from the Coast Range to the south. They planted eucalypts, an exotic from Australia. And, as easy money ebbed and flowed, they planted houses.

Settlement encouraged two trends. One filled up wildland with city; the other, cityscape with trees. Before settlement some 2 percent of Oakland is estimated to have been in woods. Between 1910 and 1913 the primary developer, Frank Havens, afforested some one and eight million eucalypts, mostly around the outskirts. The City Beautiful movement encouraged internal plantings. By the late 1950s roughly 21 percent of Oakland was treed. As the city spread outward, it absorbed more of its adjacent wildlands or, more accurately, open space. In the mid-1930s probably half of the nominal city was vacant when citizens organized a tax district to support a system of parks. By 1988 only 20 percent remained open, and that was secured under the auspices of the East Bay Regional Park District. The "wildlands" claimed the hills and their backsides. The city spread below. The two trajectories—houses and trees—crossed in 1988.[5]

What joined the two realms was the Diablo wind. What did not join them was the kind of common administrative fire service developed by the contract counties of the south, the California Department of Forestry, or the U.S. Forest Service. The East Bay thus evolved a diminutive echo of the Southern California scene, with a regional park system taking the place of national forests, but it did so without comparable fire institutions. Alameda County did not evolve along the model of the Los Angeles County Fire Department in which a single agency had to fuse fires from both wildlands and cities. Oakland remained, proudly, an urban fire department. It had the only apparatus capable of attacking a high-rise fire in the East Bay. It did not have air tankers.

As its landscape changed, so did the character of its fires. When Oakland was a wharf with back streets of shops and saloons, "recurring fires . . . almost every year swept over the hills," according to the *Oakland Tribune*. They did to the grasslands what sailors did to their ships each year when they careened and cleaned them of barnacles. As settlement pushed farther up the hills, flames and city clashed, and fires ceased to be seasonal nuisances and became historic milestones.

On September 27, 1923, a Diablo wind drove flames through the Berkeley Hills and into the campus of the University of California. Some 3,100 acres and 584 houses burned. Ten years later a Diablo-driven fire scorched a thousand acres and five homes. In 1946 another thousand acres burned. While the early Diablos identified "smokers" as a cause, the newer ones charged arsonists. Meanwhile smaller fires under the influence of westerly winds burned 10, 30, and in one exceptional instance, 700 acres. These were big numbers for a city but negligible for grassy wildlands. Then in 1970 after a hard frost had killed many blue gum eucalypts and rained litter over the landscape, a Diablo wind powered fire through 204 acres and 37 houses. It was the first outbreak in what became the statewide 1970 fire siege.

Alarm among the wildland fire community was acute. The nearby University of California, Berkeley, after all, hosted one of the premier forestry and fire science programs in the country. When drought and frost returned in 1977, the prospects for East Bay fire found its way into congressional testimony. What boosted concern was the character of the settlement along and under the summit of the hills. These were the residencies of the East Bay elite, the Northern California equivalent to Malibu or the Hollywood Hills, even if they were more likely to house Nobel laureates than movie stars. They made conflagrations the celebrity fires of the Bay Area. A small but damaging fire in 1980 kept the pot boiling. A scheme to establish a demilitarized zone between parklands and city was deemed both damaging and ineffective, but led to a Blue Ribbon Fire Prevention Committee, chaired by William Penn Mott (later director of the National Park Service), which issued its report in 1982 and recommended a fuelbreak, although tempered by aesthetic considerations. Still, much as with San Francisco prior to 1906, fire's threat remained more vivid than its reality.

Then came the 100-year burn. On October 20 the Oakland fire department knocked down a fire that started as a warming or cooking campfire amid a spot of pine near Marlborough Terrace. Crews mopped up by soaking the perimeter lines. The next day, while crews were on site and rolling up hoses, the still-smoldering hotspots disgorged embers, the flames got into patches of Monterey pine litter untouched by hose lines. The rekindled fire raced up a largely grassy slope to the ridgeline. There the Diablo wind caught it, and the fire blew up. With stunning speed it burned out the basin below Grizzly Peak. It burned through the Parkwood Apartments. It burned out the Hiller Highlands. It burned out Grandview Canyon. It burned over Highway 24. In the first hour it consumed 790 structures, each of which scattered new sources of ignition. What became known as the Tunnel fire spotted over Lake Temescal. It burned through the Rockridge District. When the Diablo winds finally slackened and northwesterly winds returned, the main front—a swarm of new ignitions, building after building—headed southeastward into Forest Park. A new index of fire spread, homes burned per hour, made its appearance. Before the orgy of burning ended, 3,354 houses and 456 apartments were ash, and 25 people had died. Total area burned amounted to 1,600 acres. It was America's worst urban fire disaster since 1906.[6]

A cataclysm this horrific scatters reports and after-action reviews like spot fires. This one sparked reviews at all levels of government, from citizen groups to the National Fire Protection Association, from a mayoral task force to California Office of Emergency Services to the National Fire Protection Association and the Federal Emergency Management Agency. As with all major disasters the surveys identified many causes, most of which had to happen together to produce results so far off the scale. Those factors that governed fire behavior fell into two general categories. One pertained to the fire environment; the other, to fire suppression capabilities.

There could be little dissent from the observation that the East Bay Hills were a prime natural setting for fire. A mediterranean climate, seasonal foehn winds, terrain that could channel fire like a coal chute, and pyrophytic vegetation that encrusted the hillside—such conditions would argue for fire anywhere they appeared. That the fire occurred amid a drought, after frosts that killed eucalypts, then a record hot spell, worsened the circumstances; yet the values were "extreme, not exceptional," and even the exotic flora only acted as an accelerant by allowing embers to kindle surface fuels and fling sparks from torching eucalypts. The "wildland" fire, however, had burned upslope toward the summit, not through the structures; and in the end, the canyon's flora survived better than its structures. The fuels that mattered were the houses and especially their wood-shingle roofs. So close were the houses that they burned one to another, and so combustible were the roofs that they both received sparks and recast them into the wind. The character of the quasi-natural setting allowed the fire to start. The character of the city allowed it to spread.[7]

The capacity to fight the fire was badly compromised. After decades of boom, Oakland went bust in the 1970s. City services decayed, among them the capacity to maintain the kind of varied fire protection demanded by the mix of landscapes within the city. Over and again, the urban fire service failed to integrate with wildland counterparts. It did not know of the red-flag warning posted by CDF for the day of the fire. It did not understand how mopup in wildland fuels differs from overhaul in buildings. It did not appreciate how a city, full of internal firewalls, might be breached from the perimeter and find itself assaulted not from the streets but from the air. It had not reckoned with fire-induced power failures that neutralized pumping stations. It had not adopted the incident command system, and could not function seamlessly with assisting agencies. It could not communicate on common radios (CDF officers resorted to telephoning dispatchers). It had three-inch hydrants, while adjacent cities used national standard two and a half. But even if compatibility had been perfect, the fire would have likely bolted away because it was moving faster than a fire department ever could. Something else intervened to break down the response.

That something was the wind. It did for Oakland what the earthquake had done for San Francisco. It simply overwhelmed the capacity to respond. The OES report noted haplessly that "a fire burning 400 or more homes per hour does not allow for normal fire-fighting tactics—either urban or wildland." Even mutual aid requires time to muster engines, planes, and personnel. In the first hour 790 homes burned. Within two hours the conflagration had reached perhaps 80 percent of its final size. The narrow streets soon clogged with traffic and fleeing residents. It was not possible to move people and cars out as fast as the fire moved in.

Converging fire engines met outgoing civilian autos. There was no Maxwell's demon in the box canyon to sort them out. There was no single flaming perimeter or high-rise to focus the action, only hundreds of individual fires—the firefight as melee.[8]

The subsequent committees, panels, boards, and task forces published hundreds of recommendations, ranked by priorities. Some involved simple changes in protocol (for example, getting daily fire weather). Many, however, required costly retrofitting, either by the city or residents in the hills (such as refitting hydrants and burying power lines). Given the parlous state of Oakland's finances, only a fraction could be enacted. But perhaps the most critical need was simply institutional: the East Bay needed the fire equivalent of its municipal utility district, or what Southern California had found with its county-CalFire-Forest Service triumvirate. The South Coast, however, had a few big, wealthy entities; the East Bay had many, smaller, and poorer ones. Even the 100,000-acre regional parklands distributed that largesse among 65 units.

Still, the reconstruction went forward. The neighborhoods rose from the ash, with better fire protection built in. After several stumbles, the East Bay Regional Park District was voted bonding authority in 2010 to expand. A Hills Emergency Forum gathered the various constituencies into a common conversation. A memorial—shaped like a gutted house with a missing roof—was erected at the intersection of Highways 13 and 23. The scars, both environmental and social, slowly healed. Ten years later the Hills Emergency Forum sponsored a review, and another 10 years later.

The threat remained dormant, not dead. A million people crowded against the hills in polyglot patches, perhaps 70 percent of them newly arrived since 1991. Parks claimed 10 percent of the landscape, a number that would rise. The municipal economy remained feeble. The Diablo still blew. Someday another Big One would shake the hills.

———————

It seems the worst expression of pedantry to fuss over labels and classifications in the face of such calamity. Yet nine days shy of 10 years after the Tunnel fire the Twin Towers burned, identifying a new urban fire threat. How that problem was defined led to a decadal war on terror that bled the country white and may have diminished rather than increased its security. Definitions matter.

What kind of fire burned the East Bay Hills in 1991? Was it an urban fire, a wildland fire, or an intermixing of the two? Did it follow a Northern or a Southern California scenario? Did it result from breakdowns in fire departments or from a social fracturing that made the ability to control fire—long considered the very essence of civilization—too difficult? Were the parklands along the hills a threat to the city or the city to the parklands? Did the fire's narrative pivot on blue gums prone to torching, or on shake-shingle roofs receptive to firebrands? What sorcerer's stew of poisons and social incantation actually stirred in the canyons' cauldron? How the problem was defined would decide what solutions were suitable. People would argue over fixes because they disagreed over causes.

Most early commentators, including myself, saw the Tunnel fire as an example of the emerging intermix fire scene. They looked at the gusty summit where wind and housing, stacked like cordwood, met, and they shuddered. They likened the catastrophe to the 1990 Painted Cave fire that swept disastrously into Santa Barbara. To their credit many East Bay parks people, often staffed with early-retired fire officers from the Forest Service, recognized the potential for fires to bolt out of open lands into the city and exercised leadership in creating something like the consortia that had evolved to the south. Many, too, had long experienced the arduous and eccentric politics of public involvement. Patiently, yet with a sense of urgency, they applied those lessons to the hills.

Yet closer inspection reveals the Tunnel fire as an urban fire. The wildlands were adjacent, but their only contribution, after a fashion, was the Diablo wind, which would blow with or without dedicated parklands. The fundamental problem was that the city had planted structures, as earlier developers had eucalypts, where they didn't belong, and then did so in ways that violated even urban fire standards. If the outcome didn't look like an urban fire, it's because the character of urban settlement had changed. San Francisco filled up its hills until the entire tip of the peninsula was built over. The Pyne clan arrived during the gold rush, and a family story holds that the patriarch once won Nob Hill in a poker game before declaring that "nothing will ever live up there but the goats" and traded the deed away for something usable (a case of whiskey). In reality, the goats were driven off, and the hills populated by flocks of wooden houses. Oakland (and sister cities like Berkeley) kept the goats. They mixed pyrophytic landscaping with the wooden houses, and had a backcountry from which fire and wind could come. San Francisco was how cities developed in the early 20th century. Oakland Hills was how they developed near the end of the century.

The prime mover is the push for high-end urban development, of which proximity to quasi-natural settings with expansive views are valued amenities. That is why the frontier between city and park exists and why quarreling is interminable about trade-offs regarding trees and houses. The story demonstrates, however, why California has the fire management system it does. Whatever the starting point, if the site is south of the San Andreas, or its East Bay offset, the Hayward, the pressures will drive the outcome to the same responses. If those measures fail, the fires will follow.

VIGNETTES OF PRIMITIVE AMERICA

The Sierra Parks

WHAT HAPPENED AMONG the giant boles is one of the fire community's origin stories. The experience did for the public lands of the West what the Tall Timbers conferences did for prescribed burning and the Southeast. The essence is that the national parks of the Sierra Nevada, notably Sequoia-Kings Canyon and Yosemite, aligned the ideas and organizations, found the determination and the luck, and mustered the right charismatic people with a charismatic megaflora to spark a new order of fire management. Among the founding generation of the fire revolution, the variants of this narrative are told and retold in meetings and memoirs, the modern equivalent of favored campfire yarns.[1]

It is a story of the right people at the right place with the right idea—an idea whose time had come. The right people: Harold Biswell, carrying the torch of prescribed fire from the Southeast to California; Starker Leopold, shrewdly identifying the removal of fire as an ecological error as devastating as the extinction of predatory wolves; a responsive cadre of rangers and superintendents like John McLaughlin and Bob Barbee; and a bevy of acolytes, mostly students like Bruce Kilgore and Dave Parsons from Bay Area universities. The roster reads like a book of prophets. The right institutions: Whitaker's Forest, an experimental field site under the direction of the University of California, Berkeley (UCB); UCB itself, the alma mater of the first two directors of the National Park Service (NPS), Stephen Mather and Horace Albright; a National Park Service, flush with Mission 66 projects, but primed for new thinking; the Sierra parks, two of the agency's crown jewels, and among the nation's sacred sites; the Leopold Committee, nominally commissioned to ponder the endless quandary of elk in Yellowstone but transfiguring that charge into a grander vision of what protected nature should be and pivoting that perspective toward the Sierras. The right ideas: the need to promote naturalness and to reinstate fire, and as the Leopold Committee found, the right expression for those notions. They said the parks should be "vignettes of Primitive America." The right timing: they bubbled up during the heady years of the 1960s, an era full of untrammeled optimism and bold innovations; the advisory board presented its findings eight months before the assassination of President Kennedy. And not least, the right setting. For a century the Big Trees had represented the ideal for shielding America's best nature from America's worst behavior. It's hard to imagine another place and time that could have brought spark and kindling together so dramatically.

Research by Richard Hartesveldt in Yosemite's Mariposa Grove, completed in 1962, and then at Grant Grove in Sequoia-Kings, argued that fire's removal threatened the Big Trees twice, once because they could not regenerate amid the overgrown clutter and again because that understory of woody litter, white fir, and cedar could carry fire to the canopies of even giants. The groves had to be burned. The Leopold Report issued the next year enlarged the domain of that observation to include the west slope of the Sierras. In 1964 Harold Biswell introduced prescribed fire to Whitaker's Forest on the flanks of Redwood Mountain. The practice crossed the boundary and made Redwood Mountain an experimentum crucis for the fire revolution. In 1967 Tall Timbers hosted its annual fire ecology conference in California, further showcasing the work of Biswell (and Harold Weaver). That winter the National Park Service overhauled its administrative philosophy.

The Green Book, as it became known, formally repudiated the 10 a.m. policy as a singular strategy for fire management. Instead, it allowed for experiments that could lead to fire's restoration. Some trials followed that summer. Sequoia-Kings ignited an 800-acre burn on Rattlesnake Ridge and allowed a lightning fire on Kennedy Ridge to free-burn. Bruce Kilgore commented dryly that there seemed no difference between the effects of those fires and that the most direct way to reintroduce fire was to stop extinguishing it. The next year Sequoia-Kings designated 129,331 acres of upper-elevation landscapes for "let-burns" and deliberately fired 6,186 acres under prescription. That year the NPS released its Green Book, with Sequoia-Kings as an implicit exemplar. The next year the burning moved onto Redwood Mountain. When, to celebrate the centennial of Yellowstone as the first national park, the agency folded a symposium on fire in the National Parks into the 1972 Tall Timbers fire ecology conference, three of the six contributions—and the first, theme-defining ones—were from Sequoia-Kings Canyon. Everglades was there of course, having authorized a prescribed fire program in 1958, but it was stuck in the far south of the Deep South, and if fire was to propagate nationally, the Sierras were where the movement would happen.[2]

The program held. It survived a flare-up on Redwood Mountain that burned hotter than desired, nomenclature that tweaked "let-burns" into "prescribed natural fires," an escape fire that required bulldozers to hold it, an organizational schism that left fire fighting with one group and fire lighting with another, and greater expenditures in costs and administration than anyone on the fireline barricades of 1968 could have imagined. The torch passed to Yosemite. But the Park Service had its proof of concept. Almost overnight the Sierra Nevada, for a while, displaced Missoula and Southern California as the hearth of western fire. The firefall over Glacier Point that Yosemite was banning as a contrived spectacle was replaced by a natural one that sent sparks cascading through the high country.

Creation stories are sudden and improbable—that's what makes them glamorous and valuable. But the power of the Sierra story is that the programs continued. They survived infancy and adolescence and grew to maturity. Once the barrier to fire's use fell nationally, experimental programs sprouted like communes. Many started, few survived.

The national narrative fixates on the Sierra parks for their generative role, the birthing of a new breed of fire. But the more interesting tale is how they grew up. This, too, can assume the form of a rebirth, and the coming-of-age story often resembles a re-creation story, and has done so for many people and places. The Sierra parks were not among them. Their story is a grittier one of internal politics, organizational tensions, lost patrons and patriarchs, public challenges and awkward controversies, missed fires and escaped opportunities, of experience trumping hope, but through it all, they persisted, they learned, they matured. In this, their overlooked saga of slogging and endurance, they may more truly represent the narrative of the fire revolution than does the oft-told tale of their inspired and audacious birth.

There is an aura of a counterculture about the early days. The fire fraternity at Sequoia-Kings was young, dismissive of the dominant community's traditional status markers, and willing to experiment. If not ready to join Ken Kesey on the bus—they looked to science rather than LSD for their visions—they were a less scruffy version of the Camp 4 compound that revolutionized rock climbing at Yosemite. They saw what they believed to be the future, and it was within the grasp of a driptorch, or better, within the withdrawn grasp of a pulaski. They would burn where they had to—amid the Mariposa, Grant, Cedar Grove, and Giant Forest Big Trees where the loss of even a single tree could be catastrophic. Elsewhere they would stay the hand itching to swat out every spark. They would let nature unburden itself of humanity's oppression.

It seemed so simple. Stop doing bad things, and good things would happen. Where you had to intervene, mimic nature. The way to get burning back into the biota was to burn. But of course nothing was simple. Between now and presettlement times when fires had free-ranged, when indigenes had burned in and around the groves and meadows almost annually, both land and society had changed in ways that made fire's reintroduction tricky. Without fire the ground had sunk under layers of needles and windfall, the woods had thickened, the fuel complex had so morphed that a fire was more likely to explode than to prune and benignly recycle. The groves would have to be mechanically thinned first; a single fire for restoration would have to yield to a series of tentative burns, each peeling back a bit more of history's ecological sediment. All this would take patience, money, institutional stamina, and a tolerance for risk.[3]

The 1972 Fire in the National Parks Symposium was a coming-out party for the fire program, and it laid down the lines of future development. Handsome, articulate, a student of Leopold's, Bruce Kilgore described a meticulous experiment in which five acres were burned to determine how fire might cleanse the understory of encroaching sugar pine and white fir. An at-times-impulsive doer, Peter ("Pyro Pete")

Schuft, chief ranger, spoke of operations. And Superintendent John McLaughlin addressed how the program might align with public sentiment and politics. The order in which the three men spoke was significant: it implied that science would inform, operations apply, and administration coordinate with the outside world.

In reality, the project was an omelet, as it had to be. What drove the program was not experimental science but the undeniable fact that fire had been a part of the Big Trees and their larger environs for millennia and was not going away. By any metric it was "natural." The choice, as Kilgore explained, "is not whether to burn or not to burn; the choice is merely when, how, and under what conditions." Schuft described an impressive escalation in projects that had put 69 percent of the park into "let-burn," had prescribe-burned 13,000 acres, and had constructed 27 miles of fuelbreaks. He judged that "we now have enough acreage burned under varying conditions to evaluate what has been done and determine methods to complete the job." With adequate funding—he thought $20,000 a year would suffice—the park could complete in five years the needed burning in critical areas and would be into "reburn cycle." McLaughlin was more circumspect. He believed the public would "wait and see," that its current attitude was a "definite plus," that the case for natural fire was clearer than for prescribed fire, and that he was "quite certain" it would be "woe to anyone who makes a mistake." The ultimate trial would come when the burning migrated into the vicinities of the General Sherman and General Grant trees. Kilgore and McLaughlin proved right, and Schuft wildly overoptimistic.[4]

The money didn't come, but the program continued. Under Barbee, with help from Biswell, an imprimatur from Harold Weaver, and research by Jan van Wagtendonk, it spread into Yosemite. The Forest Service dispatched Bob Mutch and Dave Aldrich to learn from the experience as they prepared plans to reintroduce fire into the Bitterroot-Selway Wilderness. By the time the Park Service issued NPS-19 in an attempt to put some system into the national melee of experiments, the Sierra parks were the epitome of modern fire management. Sequoia-Kings's 1979 fire plan could serve as a template for how to restore fire. The latest textbook on wildland fire, published in 1984, identified the park as a model. The use of an advisory fire committee that reported to the superintendent spread widely.

Then came the pushback. Nationally, the fire revolution stalled during the 1980s as both weather and politics polarized, funding dried up, fire science imploded and for the Park Service never really took, and the wildland-urban interface emerged as a disturbance in the Force. Locally, the galvanic personalities had retired or transferred, and their replacements were often less enthusiastic. (Bruce Kilgore temporarily left to work for the Forest Service before returning.) Among the quirks of the National Park Service as a land management agency was the segregation of fire management into two distinct entities. Fire fighting belonged under the Protection division; fire lighting, under Resource Management. Protection was the traditional career path for rangers, the agency's managerial caste. An internal schism could result when personalities, or perceived interests, clashed, as they did at Yosemite. A riot in the valley pushed the park deeply into law enforcement. Jan van Wagtendonk observed that the arrival of a new fire officer in the Ranger division "set the program back decades." Not until the 1990s, or later, did an institutionally cohesive fire program emerge.[5]

At Sequoia-Kings Canyon internal criticisms went public about the ends and means of the program in 1982. Then the 1985 Broken Arrow prescribed burn in the Giant Forest sparked controversy as Eric Barnes, a resident of Three Rivers and a former park seasonal, objected to the deep scorching and what he regarded as ideological management of an irreplaceable heritage. There was enough noise that the park commissioned a review from a committee led by Norm Christensen, a fire ecologist at Duke University, which issued its report in 1987. It affirmed that fire's restoration was necessary but that practice needed more care (and funding). Once again, Sequoia-Kings became a harbinger because the next year Yellowstone blew up, both advertising natural fire as an idea and damaging the ability of agencies to execute it. Several commissions reported on the blowout, including one by Christensen, and they recapitulated his earlier conclusions, that policy was sound but application flawed.[6]

Even when they stumbled, the fire programs at the Sierra parks commanded attention. At Sequoia-Kings the kinds of fires that elsewhere shut down programs only tempered its steel.

Yet banking coals is less exciting than starting a conflagration. After the Yellowstone crisis forced all natural fire programs to shutter until they passed reviews, the Sierra parks were among the first to reboot. That the program survived was testimony to the tenacity of the vision and the ability of its proponents to persist.

Each stumble, locally or nationally, made fire management more difficult. Checklists for approving prescribed burns grew. Funding for pretreatments or planned burns shrank. Air quality blocked major burns that would, as part of evening inversions, pour smoke into the Central Valley, and as exurbs sprouted along the east slope of the Sierras, fire officers could no longer simply shunt that smoke into Nevada. Newcomers were often less invested in a risky enterprise than founders. Pyro Pete's belief that he could turn the fire scene around in five years looked hallucinatory: it would be only marginally further along in 50 years. By the mid-1990s, as the fire revolution rebooted nationally, a program review concluded that the park—still an exemplar—was burning only about 15 percent of what the biota had known historically. The existing methods were simply not scaling up. Meanwhile, air quality in the San Joaquin was among the worst in the nation, the early warning signs of climate change were felt, and the originating assumption that the park could identify either a usable past or a desired future faded. Air quality considerations limited maximum burns to 100 acres. No one had reburned at Redwood Mountain. The 2000 National Fire Plan directed efforts into fuels, not ecosystems.

Still, compared to most of the country, the Sierra parks remained stellar programs. Some 97 percent of Kings Canyon operated under a natural fire agenda. Several wildfires had threatened groves and were quelled thanks to earlier treatments. The administrative division of fire management, like Rome's armies governed by two tribunes, each ruling for a day, was resolved by a merger. The fire history of Yosemite's Illilouette Creek Basin established itself as a textbook example of natural fire that, within limits, became self-regulating. At the Sierra parks it would be a gross misstatement to declare the revolution a failure because it did not satisfy its utopian goals. By proving an alternative to suppression was possible, the Sierra experience made the revolution, however compromised, plausible.[7]

But it was clear that the house odds were against fire's rational introduction. Every act and decision was a fight: the fire program could only spread if it was continually reignited amid a tricky fuel array and oft-skeptical public scrutiny. The Green Book and its successor federal fire policies allowed for fire's restoration but did not mandate it. Each year the program fell a little further behind. The only burning at scale was wildfire. Yet without the exhaustive tending its devotees lavished, the program would have expired.

If the revolution would come anywhere outside Florida, the Sierra parks, particularly Sequoia, were the likely candidates. Even when the army ran it, prior to the creation of the National Park Service in 1916, fire protection meant mostly quelling the burning that locals and ranchers did deliberately and tourists accidentally. The Park Service assumed the programs of the cavalry as it did their uniforms, but staffing fell far below what would be necessary for serious suppression. Besides, Colonel John White, Sequoia's charismatic superintendent, believed in light-burning, at least around the Big Trees.

Then a large fire in 1921 embarrassed the park, but while plans to bolster its fire program floated in its wake, they lacked the funding to implement. In 1926, after a disastrous season at Glacier National Park, the NPS got serious. In 1928 it hired a national fire control officer, John Coffman, from the Mendocino National Forest and secured line-item funding from Congress for fire protection. Still, a few miles of trail, the occasional fire cache, and a solitary lookout were hardly adequate. The big push came with the CCC, though most of its energies went to major visitor sites and the Big Trees. Mission 66, a 10-year investment in national infrastructure begun in 1956, revived the CCC project. The fire program got attention following a major wildfire at Tunnel Rock in 1960 and a scare when the 1961 Harlow fire raced outside the boundary. The outcome was a helitack operation stationed at Ash Mountain in 1962. But that was the year the Leopold Committee was commissioned: the buildup had barely begun before it was challenged.

In the end, while the National Park Service nationally, and Sequoia-Kings Canyon specifically, had accepted suppression as official policy, it had never had the money or heart to apply the doctrine rigorously. Much of the High Sierra backcountry had never been seriously affected, save for removing the shepherds and other vagrants who had kindled so many fires during their annual transhumance and, at Yosemite, had inflamed the indignation of John Muir. And there remained a sullen if quiet defiance whose genealogy traced back to Colonel White.

The parks, and the groves specifically, were not in good shape. But the lands had not been scalped by logging and overgrazing or had their fundamental structure upended as had occurred in so many reserves. It seemed self-evident that, if removing fire had allowed nature's ebullience to overflow the scene, then restoring fire could, by itself, correct the imbalance. By the mid-1960s the old undercurrent of resistance at Sequoia-Kings had strengthened into a riptide. The Sierra parks wrote the red book for America's great cultural revolution on fire.[8]

An expectation was in the air that the fires in the hallowed groves would become a point of positive infection, like ancient fire rituals in which stale fires would be extinguished and fresh ones brought to the countryside from new-kindled sacred one. That didn't happen. If the feudal structure of the national park system encouraged experimentation, it also stalled efforts to spread systemically. Fire programs propagated less by institutional mechanisms than by people, like Bob Barbee taking Yosemite's model to Yellowstone. Or not. Meanwhile, the headwinds of the outside world drove the flames back onto themselves. By the new millennium the environmental (and political) circumstances that allowed wildfires to metastasize into megafires also made fire's management impossibly cumbersome. The final 2004 fire plan (and environmental impact statement) for Yosemite ran to 1,878 pages, stood two inches high, and weighed 4.1 pounds. It was the administrative equivalent of 1,000-hour fuels.

There was little means to force reform, and as blowouts like the 2000 Cerro Grande escape fire punished the people responsible, there were concerns about the project that looked a lot like fear. Yet Sequoia-Kings, in particular, remained a refuge and rally point; its idea, and frequently its administrative models such as its fire committee, traveled with transferred personnel and rooted elsewhere in the national park system. The deeper lesson was the limits to the spread of fire restoration. The revolution was not self-perpetuating. It had to find, or be fed, new fuels. If it didn't grow, it would expire.

Yet it would be wrong to conclude that the experiment failed. It had worked. Both Yosemite and Sequoia-Kings Canyon cultivated fire programs that remained the envy of the Park Service and served as paragons nationally. A large fraction of NPS fire leaders came out of, or did a tour at, the Sierra parks. But paradoxically the parks' purpose had only become national by looking to their own lands first, and they continue to measure requests from the outside agencies or the California fire plan against the internal requirements of their program. The national or regional priorities for fire do not drive their operation; their local needs do. They do not wear fire bugles on their lapels. They keep the traditional sequoia cones.

As always it comes back to the Big Trees.

While popular legend holds that the first national park was Yellowstone in 1872, cognoscenti know that the original national project to protect a natural spectacle happened in 1864 when Congress reserved the Mariposa Grove of Big Trees to keep them from being logged, and threw in Yosemite Valley to spice the pot. Then the Feds ceded the reserve to the state of California, and when California failed to provide proper protection, it reclaimed the land in 1890 to become Yosemite National Park; at the same time, Sequoia and General Grant National Parks were authorized to protect other groves; in 1940 Kings Canyon joined them; and over time new landscapes like Tehipite Valley and Mineral King were added, and the parks were merged or administered out of a common center. The power of the Big Trees brought other landscapes in tow. That history is emblazoned on the arrowhead patch of the National Park Service and on the sequoia cones that decorate its official Stetson and lapels. The park idea originated with the Big Trees; it then propagated by example rather than by direct lineage.

So it has been with fire management. The fire revolution in the American West began with sequoias and carried the backcountry with it, some of it far from the Range of Light. But a national campaign was not the point: the historic role of the Sierra parks was less to spearhead the invasion of an idea across the public domain than to restore fire to the sequoia sanctuaries. They were not beachheads for a new idea so much as its rally points and retreats. Like the United States in the 1860s, the National Park Service lacked an institutional mechanism by which to propagate a radical practice throughout the country. Instead, the Sierra parks grew organically, absorbing the mountains beyond the groves. However much their example might inspire others, the parks looked first to themselves. They hold fire as they do the Big Trees.

EPILOGUE

California Between Two Fires

Here the granite flanks are scarred with ancient fire
. . . .
Beautiful country burn again, Point Pinos down to
the Sur Rivers
Burn as before with bitter wonders. . . .
—ROBINSON JEFFERS, "APOLOGY FOR BAD DREAMS"

Fire is an old story.
I would like,
with a sense of helpful order,
with respect for laws
of nature,
to help my land
with a burn, a hot clean
burn.
. . . .
And then
it would be more
like,
when it belonged to the Indians.

Before.
—GARY SNYDER, "CONTROL BURN"

IN 1964 CARL WILSON, whose Forest Service career had spanned both research and practice and made him probably the most knowledgeable fire authority in California, announced the obvious: "It is hard to describe the problems in California without using superlatives." Since the 1930s, he noted, everything relevant to California fire had bulked up by a factor of two to ten, and even as he wrote the state was at a tipping point. A year earlier the Leopold Report had been released and the Riverside Fire Lab had opened. In the Sierras field trials were underway to reintroduce fire to the Big Trees, while along the Transverse experiments continued into mass fire and conflagration control. The split between the two Californias was widening, and the impact of California nationally was amplifying.[1]

Nothing in the past 50 years since Wilson wrote has lessened either trend. Of the three practices that emerged from the fire revolution of the Sixties to define the national fire scene—prescribed fire, natural fire, and fire suppression along the intermix—California experimented with them all and defined the terms of the last two. That impact continues. That what happens in California will happen to the United States is not just metaphoric hyperbole: the state's population and economy assure it will shape national policy, and beyond that heft, its capacity for media hype will leverage its behavior into the national consciousness.

There are plenty of examples of how this works. Water management is one, as California's unilateral decisions regarding a state water plan captured federal agencies, which distorted regional and even national policy. Perhaps even more relevant is smog control. Its peculiarities—geographic and social—made smog worse in Los Angeles than elsewhere, which caused it to set special standards for abatement, which then migrated to the state level, which, because California's standards were higher than those nationally and its market for automobiles commanding, affected national norms. That, in brief, is what has happened with fire.

Its defining divides persist. The border between north and south remains distinctive. Each region has its characteristic fires and charismatic fire species, even its chosen poets—the enlightening flames of a Gary Snyder, the apocalyptic fires of a Robinson Jeffers. The two parts rub along like the San Andreas, with most of the institutional tension absorbed invisibly, but occasionally broken by massive rupture.

That geographic border, however, masks a deeper divide that increasingly informs the state's fire history. It is the pyric transition that moved anthropogenic fire from open flame to closed combustion, most spectacularly through the internal combustion engine (ICE). Like the nation generally, but with more intensity and visibility, California has transitioned from fire to ICE.

Two fatality fires nicely highlight that shift. Investigating the Loop fire disaster of 1966, Bud Moore wrote in his journal that Bill Derr drove him through a rainstorm past "four serious accidents on the freeway, each of the wrecks involved 3 to 4 cars. This shook Bill not at all." The Loop fire, however, killed 11 members of the El Cariso Hotshots and became the subject of a national inquiry and a gut-wrenching review of fireline safety. The Foothill freeway accidents may have killed as many, but could be overlooked, even by those obsessed with the Loop catastrophe. When flames took lives, it was an event beyond the quotidian world and commanded attention. When ICE killed, it was, however regrettably, the way the world worked and could be safely quarantined from the imagination. When, in 2006, a Forest Service engine crew perished in the Esperanza fire while defending structures, the episode sent shockwaves through the California fire community, and the arsonist responsible for setting the blaze received the death penalty. When, two years later, a Sikorsky S-61 ferrying fire crews on the Iron Complex in the Shasta-Trinity National Forest crashed, killing nine and injuring four others, they were mourned and honored as fallen firefighters, but there was no further action other than to cite Carson Helicopters for safety violations. The tragedy seemed to lie outside the purview of fire management (and perhaps reviewers did not wish to peer too deeply since the fire burned in wilderness, not far from a 200,000-acre megafire from two years previously into which the Iron Complex would have burned). They were felled by ICE, not fire. More firefighters die now in machines than on firelines.

What makes the contrast between the two fires doubly intriguing is that it crosses the grain of California's north-south divide, or even its fractal borders between the wild and the urban. The irony is palpable: Southern California, famous for its car culture, had firefighters die in flaming chaparral, while Northern California, in this case the Trinity Alps Wilderness, lost firefighters to engine failure. The two fires, open and closed, are inscribing their own geography and dynamic over the state. Unlike those other frontiers, however, this one has no science or policy behind it. Its effects are not considered the subjects of ecology. Its pyrogeography is not reckoned as informing. Yet this divide will continue to impose itself over the others, and the transition it depicts will likely determine the future of California fire and, through California's capacity to exaggerate and export its obsessions, the evolution of fire management in America.

In a sense the transition offers a parallel geography to the Transverse. To the east, at San Gorgonio Pass, where the Big Kink begins its pivot westward, the wind quickens and concentrates, like a sluggish river suddenly forced through a narrow gorge. The outside portal is awash with wind turbines—the very emblem of sustainable energy and an important factor in California's determination to wean itself off fossil fuels. To the west, just beyond the Santa Monica pivot where the Transverse Range turns northward, offshore oil rigs dapple the Santa Barbara Channel, the epitome of California's second great mineral rush and the motive combustion behind its industrialization. In 1969 oil spillage helped kindle a local environmental movement, which then fed into Earth Day. Between those two points stretch the mountains and the city, the wild and the urban, the free-burning and the fuel-injected. Between their fires lies the Big Kink of America's future with fire.

───────

During its postwar heyday, California furnished two two-term presidents. One left in political scandal and a ruined economy. The other skated over scandals but silently trashed the economy through structural deficits, or what Office of Management and Budget Director David Stockton prophesied as "red ink as far as the eye can see." When it could no longer fail on the national scene, California turned inward. By the early 1990s, without the federal subsidies from the Cold War, California was imploding in almost every way, and a natural disorder matched the social. Mike Davis sardonically termed the scene a "theme park of the Apocalypse."[2]

Earthquakes, floods, and fires; gangs, immigration, and the collapse of a middle class—all seemed insoluble. The state sank into insolvency. Money flowed from emergencies, or was siphoned into prisons and casinos. San Bernardino went from being ranked in 1976 as one of the most desirable cities in the United States to becoming one of its bottom-feeders 20 years later, as *Money* magazine downgraded its civic qualities to the status of junk bonds and declared it the worst city in the nation in which to raise a family. Social cohesion, it seemed, had to appeal to security services, and both the Los Angeles Police Department and the fire services turned to state-of-the-art machinery and the application of overwhelming force.

The economy foundered. It was built on sand and serfs—the sand that went to make concrete for construction, the sand smelted into silicon for the digital revolution, the de facto serfs who provided a mobile, quiescent workforce for farms, factories, and domestic service. Both sand economies rose on phony money and "irrational exuberance." The dotcom bubble burst in 1999, and the subprime housing bubble in 2007. The workforce relied on an influx of immigrants, many illegal. Without easy money to cover overpromised IOUs, the partisan politics turned toxic: the state seemed politically paralyzed, able only to borrow or finesse around insolvency. During the gold rush California had helped finance the Union. Now, the nation floated California. By 2009 even national magazines like *Time* questioned how California could fund its traditional firefights, for fire could not stand apart from the rest of the state—how could it?[3]

One of California's native poets, Robert Haas, later poet laureate of the United States, once observed that the state's history might be imagined in fact as a succession of fires (he thought four). The idea that fire could serve as an organizing conceit for the California experience came easily to the mind of any serious observer, since the state's history is often a chronicle of social and natural cataclysms happening in parallel. In 2003, a wave of conflagrations accompanied the political turmoil that culminated in a gubernatorial recall election that replaced Gray Davis with Arnold Schwarzenegger while the state struggled to rebuild after the dot-com bubble; and other sieges, even larger, struck both south (in 2007) and north (2008) as the state imploded with the subprime mortgage crash and shuffled into political paralysis and fiscal insolvency. The only response was to mobilize ever greater forces paid with emergency money. The sand economy melted, or washed away in the postburn floods.[4]

The rest of the country might snicker or savor a scent of schadenfreude at the spectacle of California's serial meltdowns. But California's fire failures would affect them as surely as its successes. California's dysfunction was not the fire leadership the country at large needed. There were successes in the Sierra Nevada and along the backside of the Big

Kink; but they were niche exceptions, not national exemplars. They said, if sotto voce, that California had not corrected the structural problems of land use, population growth, public financing, and infrastructure decay that had caused the state to sag. So its fire scene mirrored its social order generally. Pressures drove fire management toward more suppression, more technology, more costs, and more public theater. The state, it seemed, faced a future of pyric riots and emergency lock-downs. Fire protection was on its way to joining other agents of a constabulary committed to stabilizing an inconstant social order.

Yet California had promise, too. It had been sold short before; more than once it has returned from the near dead. Despite its enormous population, much of its land is still open, and despite the modern incarnation of the governing class into a new plutocracy, so is its society; they will remain so. Even Southern California has ample room to continue what it has done for most of a century. In brief, there are plenty of fires and landscapes on which to manage them; there are abundant opportunities to experiment and reinvent; and the heft of its statewide fire establishment means that what one part needs another can usually furnish.

What California requires is what, in principle and myth, it has always promoted: a new start. Its politics needs a fresh constitution, and its fire scene a radical matrix for joining its parts into a working whole. It needs on the level of the state commonwealth what the National Cohesive Strategy, mandated by the 2009 FLAME Act, proposes for the country overall, a way of determining rights, roles, and responsibilities that does not force everything into a common sluice of suppression, that does not impose firefighting as a default setting nor depend on a torrent of federal subsidies to keep engines pumping and air tankers flying. California as a state-nation has too quickly inflated to a higher governmental level what might better be handled at a lower one. The reason for this escalation of course is the state's (and especially Southern California's) proneness for conflagrations. But cataclysm is not a sustainable formula for fire management. Ultimately, it's a question of politics more than policies. Disaster response makes for good TV but bad practice. Not least, its tremblor-prone fire economy makes California a suspect anchor point for managing fire nationally.

California and America need to renegotiate their fire-mediated relationship. California needs to recognize that it exerts national leadership whether it deserves it or not, or even whether it wants it or not, and that national leadership brings costs, not merely perks. It cannot be both leach and leader, although it will shape the national agenda either way. Rather, it needs first to look to its own house, let fire find its own level rather than moving engines around the state as it does water, and accept the limits of what it can accomplish with the means it possesses and not pretend that other gold rushes (virtual or otherwise) will compensate. Equally, the country must reciprocate. California is too big to ignore, too volatile to quarantine, and too eccentric to allow it to establish national norms by default. The nation must find ways to preserve what the endless experiment that is California has churned up that is most valuable beyond its borders and flense away the rest. Much of what is Californian doesn't work even in California. California is a part of the whole; it is not a part that speaks for the whole.

During the fire revolution, California designed a system to replace what it (and much of the country) found objectionable. The times they were a-changin'. But then they kept on changing, in California more than elsewhere. Like much that bubbled out of the cultural cauldron that was the Sixties, the reforms proved inadequate as administrative prescriptions. What began as a bold, even utopian bid to prospect a new world of fire is ending, half a century later, with an ever-more-costly scramble to protect the indefensible consequences of what in fact resulted.

Now Golden California finds itself caught between the same fires as much as the rest of the country. Only more so.

NOTE ON SOURCES

A BIBLIOGRAPHY OF CALIFORNIA FIRE could be a book in itself. Mostly I have limited my references in the notes to citations for quotations or particular facts. But it may be useful to collect some of the basic sources here for readers interested in background.

Begin with fire, and with the encyclopedic survey offered by Neil G. Sugihara et al., eds., *Fire in California's Ecosystems* (Berkeley: University of California Press, 2006), which includes fire history as well as ecology. A handy distillation is available in David Carle, *Introduction to Fire in California* (Berkeley: University of California Press, 2008). For a delightful memoir from a pivotal researcher, see Harold H. Biswell, *Prescribed Burning in California Wildlands Vegetation Management* (Berkeley: University of California Press, 1989).

On human fire history, there is M. Kat Anderson, *Tending the Wild: Native American Knowledge and the Management of California's Natural Resources* (Berkeley: University of California Press, 2005), which amplifies the pioneering study by Henry T. Lewis, *Patterns of Indian Burning in California: Ecology and Ethnohistory* (Ramona, CA: Ballena Press, 1973). The state is particularly fortunate in having good records and histories for its Board of Forestry, as it evolved into its current avatar, CalFire. See C. Raymond Clar, *California Government and Forestry from Spanish Days Until the Creation of the Department of Natural Resources in 1927* (Sacramento, CA: Division of Forestry, 1959) and *California Government and Forestry - II: During the Young and Rolph Administrations* (Sacramento,

CA: Division of Forestry, 1969). CalFire has a very handy repository of the major commissions, reports on big fires, and so on in its offices, apart from the state archives. Two other helpful books on land history are Samuel Trask Dana and Myron Krueger, *California Lands: Ownership, Use, and Management* (Washington, DC: American Forestry Association, 1958) and Forest and Rangeland Resources Assessment Program, *California's Forests and Rangelands: Growing Conflict Over Changing Uses* (Sacramento: California Department of Forestry and Fire Protection, 1988).

The federal story is largely that of the U.S. Forest Service. Again, the agency has itself authored its own narrative. See Robert W. Cermak, *Fire in the Forest: A History of Forest Fire Control on the National Forests of California, 1898–1956* (USDA Forest Service, R5-FR-003, 2005); Anthony Godfrey, *The Ever-Changing View: A History of the National Forests in California* (USDA Forest Service, R5-FR-004, 2005); and Victor Geraci, comp., *The Lure of the Forest: Oral Histories from the National Forests in California* (USDA Forest Service, R5-FR-005, 2005). For the National Park Service, consult the various histories of the parks that are available and Hal K. Rothman, *Blazing Heritage: A History of Wildland Fire in the National Parks* (New York: Oxford University Press, 2007), which includes a nice accounting of the Sierra parks and the fire revolution.

And finally, on the matter of California as an entity and idea, I have relied on several classics: Carey McWilliams, *California: The Great Exception* (1949; repr., Berkeley: University of California Press, 1999) and Kevin Starr's Americans and the California Dream series, particularly *Coast of Dreams: California on the Edge, 1990–2003* (New York: Vintage Books, 2006). A tidy distillation is available in Starr, *California: A History* (New York: Modern Library, 2007). Joan Didion's *Where I Was From* (New York: Knopf, 2003) suggested I might find a foil to the usual data-driven science and policy publications by thinking of literary referents.

While all these publications were informative in my research, and a couple inspirational, *California: A Fire Survey* did not derive from them but from my on-the-ground travels and the need to reconcile words with what I saw and heard.

NOTES

CALIFORNIA'S INVENTED FIRE CULTURES

1. Josiah Royce, *California: A Study of American Character* (1886; repr., Berkeley: Heyday Books, 2002), 4.
2. Royce, *California*, 307, 304.
3. Joaquin Miller, Proceedings of the American Forestry Congress at Its Sessions Held at Cincinnati, Ohio in April 1882 and at Montreal, Canada in August, 1882, at Its Meeting Held in Boston, September, 1885 (n.p.: American Forestry Association, 1886), 26.
4. The most thorough description is M. Kat Anderson, *Tending the Wild: Native American Knowledge and the Management of California's Natural Resources* (Berkeley: University of California Press, 2005). It much amplifies the pioneering study by Henry T. Lewis, *Patterns of Indian Burning in California: Ecology and Ethnohistory* (Ramona, CA: Ballena Press, 1973).
5. Chuck Mansfield, letter to author, August 4, 2011. The episode occurred on the Malheur National Forest, in Oregon, but was relatively common throughout the West in that time.
6. Ron Watson, letter to author.
7. Roderick Nash, *Wilderness and the American Mind*, 5th ed. (New Haven: Yale University Press, 2014).
8. Propagation is not even restricted to the United States. In an effort to beef up its fire protection capabilities, Mexico is receiving training from

CalFire. As one skeptic notes, this means the Sierra San Pedro Mártir "will go the way of California forests." Baja California, indeed.

STATE OF EMERGENCY

1. Compared with most agencies CalFire has successfully preserved a record of its past. The agency generously made available that in-house archive to me. I wish to thank Janet Upton for setting up my visit and Mary Welna for introducing me to that rich storehouse. I also wish to thank Chief Ken Pimlott for an opportunity to meet. Kim Zagaris of CalEMA went even further, not only creating a most useful interview with Richard Barrows but handing over a cache of copied documents. I regret, as so often, that an essay on a topic that the agencies themselves may regard as tangential is the outcome to what could easily constitute a book in itself (and in CalFire's case did, published for its centennial). My task is to place the agencies within California and California within the national fire scene. My thanks to all.

2. California is exceptionally fortunate to have an extensive history of state forestry operations in the form of Raymond Clar's two-volume opus, *California Government and Forestry* (Sacramento, CA: Division of Forestry, Department of Natural Resources, 1959, 1967). Abridgements are available in C. Raymond Clar, *Evolution of California's Wildland Fire Protection System* (Sacramento: California Division of Forestry, 1969) and Mark V. Thornton, "History of CDF," in California Department of Forestry and Fire Protection, *100 Years of CDF* (Paducah, KY: Turner, 2005), 10–20.

3. Clar, *Evolution of California's Wildland Fire Protection System*, 23.

4. Clinton B. Phillips, *California Aflame! September 22–October 4, 1970* (Sacramento: California Division of Forestry, 1971).

5. Task Force on California's Wildland Fire Problem, *Recommendations to Solve California's Wildland Fire Problem*, submitted to California Resources Agency, June 1972.

6. Many good accounts of Firescope exist. The core remains Richard A. Chase, *FIRESCOPE: A New Concept in Multiagency Fire Suppression Coordination*, U.S. Forest Service, General Technical Report PSW-40, 1980. Supplement and update with the following documents from CalEMA: *Some Highlights of the Evolution of the Incident Command System as Developed by FIRESCOPE*, unpublished timeline; *FIRESCOPE California: Past, Current and Future Directions: A Progress Report* (October 1988); and *FIRESCOPE's Future*. Useful for a more personal

perspective is the treatment of FIRESCOPE in Region 5 Oral History Project, *The Unmarked Trail: Managing National Forests in a Turbulent Era: Region 5 Oral History, Volume II: 1960s to 1990s* ([Vallejo, CA?]: U.S. Forest Service, Pacific Southwest Region, 2009), 153–87.

7. An excellent chronicle is available in CalEMA, *California Fire Service and Rescue Emergency Mutual Aid System: History and Organization* (rev. April 2002).

8. Information from California Department of Forestry and Fire Protection, *100 Years of CDF*, especially 21–23. The book makes a case that the change from green to blue pants was required because the old pants, which included polyester, were a fire hazard, and blue was the only market alternative. But the federal agencies had developed green Nomex pants. The choice was clearly made to align more closely with urban counterparts.

9. Ibid., 31.

FOUR FORESTS

1. For my tutorial on the San Bernardino, I wish to thank John Miller, David Kelly, Rocky Opliger, Dan Snow, Scott Wagner, and Susan Zahn who contributed data, ideas, and that most opulent of gifts, their time.

2. For my introduction to the Angeles, I benefited enormously from a morning's discussion with Mike Rogers, Don Garwood, and Jack Lane, organized by Diane Travis, who also supplied some background data. The group had probably a century of on-the-ground fire experience in the San Gabriels. I'm indebted to all of them for their insights. It should go without saying, however, that my treatment brushes against only a tiny edge of the Angeles fire story. Nor should anyone assume that my choice of topic and treatment agrees with theirs.

3. Clive M. Countryman, Morris H. McCutchan, and Bill C. Ryan, *Fire Weather and Fire Behavior at the 1968 Canyon Fire*, U.S.D.A. Forest Service Research Paper PSW-55/1969, Pacific Southwest Forest and Range Experiment Station, 1969, 17–18.

4. U.S. Forest Service, Fire and Aviation Management, *Station Fire Initial Attack Review: Report of the Review Panel*, November 13, 2009, 7–10.

5. Rule 409.5 is state law, but the federal agencies accept it to reduce confusion and maintain an interagency front for fire protection.

6. For a primer on the Cleveland, I would like to thank John Truett, Clay Howe, Anne Fege, and especially Stephen Fillmore for introducing me

to the Cleveland. We spoke little about ignition; in truth, the statistics were appallingly poor. But they gave me a sense of the setting in which ignition would occur, which allowed me to bring the forest into the conceptual design of the fire rectangle. I admire their passion for the place and am grateful for their willingness to pass along a portion of their hard-won wisdoms.

7. For my introduction to the Los Padres, I'd like to thank Andrew Madsen for allowing me to join a field trip to the scene of the La Brea fire, Chris Childers for a terrific tutorial on the fire and the Santa Barbara County fire scene generally, Celine Moomey for maps and data, and Anthony Escobar for his insights into fire on the Los Padres. I regret my abstract of an essay cannot give full honor to their understanding and assistance.

THREE PARKS

1. Special thanks to Marti Witter for catalyzing a most informative tutorial with SMMNRA staff. No outfit had so much information in a usable form, and none was more generous with what they had. The resulting essay is a molehill birthed from this mountain. In terms of its density, it's an embarrassment. I can only hope the novel arrangement of parks into a transect can create context that might add some value to the library.

2. For a useful introduction, see U.S. Department of the Interior, National Park Service, *Joshua Tree National Park: Fire Management Plan (Twenty-nine Palms, California): Environmental Assessment*, April 2005. The park's fire management website has links to the critical published science.

3. Sources: Channel Islands National Park, *Wildland Fire Management Plan* (2006), http://www.nps.gov/chis/learn/management/upload/CHISFMP_FinalVersion_6_1_2006%204.pdf; "Prospectus: Ecological Fire Program Needs at Channel Islands National Park" (USGS-BRD, Channel Islands Field Station, and Resource Management Division, Channel Islands National Park, June 14, 1999); Elizabeth Hobbs, "Vegetation Dynamics of a California Island," 603–6, and Richard A. Minnich, "Grazing, Fire, and the Management of Vegetation on Santa Catalina Island, California," 444–49, in C. E. Conrad and W. C. Oechel, eds., *Proceedings of the Symposium on the Dynamics and Management of Mediterranean-Type Ecosystems*, General Technical Report PSW-58, U.S. Forest Service, 1982.; Santa Barbara Museum of Natural History,

California Islands Symposia (three to date); and *Channel Islands Fire History Summary*, Channel Islands Fire Ecology Workshop, March 1, 2002.

4. Ishmael Messer, Wildland Fire Management Plan: 1994 Revision: An Addendum to the Resource Management Plan for Santa Monica Mountains National Recreation Area, Santa Monica Mountains National Recreation Area, 1994, 47-50, and appended comments by Steve Bakken, July 8, 2002.

IMPERIUM IN IMPERIO

1. John Todd and J. Lopez granted several hours of their busy schedule to present the labors of their centennial celebration, to coach me in the scope of LACFD operations, and to escort me around the facility at Pacoima. It was an enlightening education. I regret my sketch can only scratch at the surface but hope those markings can hint at the essence beneath.

2. An encyclopedic survey of LACFD up through the late 1980s is available in David Boucher, *Ride the Devil Wind: A History of the Los Angeles County Forester and Fire Warden Department and Fire Protection Districts* (n.p.: Fire Publications, Inc., 1991). I have mined the text for the basics, supplemented by other documents and discussions as indicated.

MENDING FIREWALLS

1. Special thanks to Anne Fege for organizing a lively group discussion over an even more delicious dinner, and to Stephen Fillmore of the Cleveland National Forest for thoughtful commentary and maps on the regional fuelbreak network, and to Clay Howe for patiently explaining the Establishment position. I'm sure they all would wish that I would get off the fence, but the endurance of the fence itself may be the story, or so it appears to a historian of my temperament.

2. As Stephen Fillmore noted in comments to a draft, "It of course is difficult to implement landscape scale management when you don't have a landscape scale to work with."

3. "International Fuel Break," California Fire Alliance, Success Stories, handout; quote on burnout from Clay Howe.

4. John McPhee, *The Control of Nature* (New York: Farrar, Straus, and Giroux, 1990), 271.

FORCE MAJEURE

1. See "Fact Sheet: Vandenberg AFB: An Historical Overview," Vandenberg AFB History Office. For general information about fire regimes, see "Wildland Fire Management Plan, Vandenberg Air Force Base, California, Sept 5, 2008." This document is now under revision. Other fundamental documents include University of California Department of Forestry & Resource Management and U.S. Forest Service, "Wildland Fuel Management for Vandenberg Air Force Base (Vandenberg, California), 1980"; Diana E. Hickson, *History of Wildland Fires on Vandenberg Air Force Base, California*, NASA Technical Memorandum 100983, National Aeronautics and Space Administration, John F. Kennedy Space Center, March 1988; and Paul A. Schmalzer, Diana E. Hickson, and C. Ross Hinkle, *Vegetation Studies on Vandenberg Air Force Base, California*, NASA Technical Memorandum 100985, National Aeronautics and Space Administration, John F. Kennedy Space Center, March 1988.

   I wish to thank Jessie Hendricks and Jim McLean for their generous conversations, Kristen Halbeisen for a tour of the base, and Dan Ardoin for helping set up a visit. All opinions expressed in this essay are of course mine, not theirs.

2. Joseph N. Valencia has performed a valuable service by assembling the story of the disaster into a single volume, *Beyond Tranquillon Ridge* (Bloomington, IN: AuthorHouse, 2004). My account derives from his, supplemented by an unpublished summary by Chief Master Sergeant Orval Given, 4392nd Civil Engineering Squadron, Fire Protection Branch, Vandenberg Air Force Base, California, among documents in the possession of the Vandenberg Hotshots.

3. Quote from Ronald F. Lockmann, *Guarding the Forests of Southern California* (Glendale, CA: Arthur F. Clark, 1981), 109.

AIRING DIFFERENCES

1. Sources. The SCAQMD relies on its web page rather than print to communicate its message and mission. The department answered a routine request for data, though not as fully or exactly as I had wished. A popular summary of air quality and institutional response is available in Chip Jacobs and William J. Kelly, *Smogtown: The Lung-Burning History of Pollution in Los Angeles* (Woodstock, NY: Overlook Press, 2008);

quote from page 10. I also found helpful passages in Kevin Starr's *Coast of Dreams: California on the Edge, 1990–2003* (New York: Vintage Books, 2004), particularly those that traced the demonization of the automobile. Finally I would like to acknowledge a too-brief if generous conversation with David DeBoer of SCAQMD.

INTERLUDE: CALIFORNIA SPLIT

1. The foundation study—a classic of fire research—is Clive M. Countryman and Charles W. Philpot, *Physical Characteristics of Chamise as a Wildland Fuel*, U.S. Forest Service Research Paper PSW-66, 1970.

2. Still regarded as Scripture is the marvelous anthology of that early research, H. Thomas Harvey, Howard S. Shellhammer, and Ronald E. Stecker, *Giant Sequoia Ecology: Fire and Reproduction*. Scientific Monograph Series No. 12, National Park Service, 1980. The original research was completed in 1970, the manuscript submitted in 1971, and published formally in 1975.

3. The best summary is Clinton B. Phillips and Jerry Reinecker, *The Fire Siege of 1987: Lightning Fires Devastate the Forests of California*, California Department of Forestry, 1988.

4. Robert E. Martin, "The 1993 Southern California Fires," *International Forest Fire News* 10 (January 1994): 21–22. The most memorable report, however, remains Mike Davis's portrayal of pandemonium in *The Ecology of Fear* (New York: Metropolitan Books, 1998).

5. California Department of Forestry and U.S. Forest Service, *California Fire Siege 2003: The Story*, California Fire Alliance, 2004; CalFire, U.S. Forest Service, and CalEMA, *California Fire Siege 2007: An Overview*, CalFire, CalEMA, USFS, 2008.

6. On the Siege's statistics, see CalFire, *2008 June Fire Siege*, 2008; on largest fires, page 14. See also Jonetta T. Holt et al., *Initial Impressions from the Northern California 2008 Lightning Siege: A Report by a Wildland Fire Lessons Learned Center Information Collection Team*, Wildland Fire Lessons Learned Center, 2009.

ARCH ROCK

1. Yosemite claims a literature suitable to its standing. I used Alfred Runte, *Yosemite: The Embattled Wilderness* (Lincoln: University of Nebraska Press, 1990) as a point of entry, and Lary M. Dilsaver and William C.

Tweed, *Challenge of the Big Trees: A Resource History of Sequoia and Kings Canyon National Parks* (Three Rivers, CA: Sequoia Natural History Association, 1990) as a guide to the role of the giant sequoia.

THE TAHOE CRUCIBLE

1. I would like to thank the Lake Tahoe Management Unit's fire staff—Randy Striplin, Cheva Heck, John Washington, Beth Brady, and Kit Bailey—for an informative introduction to fire in their part of the world and for helping me access an unusually robust body of studies. For what might be insightful in my abbreviated survey, they can take credit. For what is wrong, I must accept the blame.

2. For general environmental history, see Douglas Strong, *Tahoe: An Environmental History* (Lincoln: University of Nebraska Press, 1984). Much richer, however, is chapter two, "A Contextual Overview of Human Land Use and Environmental Conditions," in Dennis D. Murphy and Christopher M. Knopp, eds., *The Lake Tahoe Watershed Assessment: Volume I*, General Technical Report PSW-GTR-175, U.S. Forest Service, 2000.

3. Leiberg quote from Murphy and Knopp, *Lake Tahoe Watershed Assessment*, 83.

4. Figures from U.S. Forest Service et al., *Lake Tahoe Basin: Multi-Jurisdictional Fuel Reduction and Wildfire Prevention Strategy*, U.S. Forest Service, 2007, accessed September 30, 2015, http://www.fs.usda.gov/Internet/FSE_DOCUMENTS/fsm9_046334.pdf.

5. U.S. Forest Service, Lake Tahoe Basin Management Unit: Kathy Murphy, Tim Rich, and Tim Sexton, *An Assessment of Fuel Treatment Effects on Fire Behavior, Suppression Effectiveness, and Structure Ignition on the Angora Fire*, U.S. Forest Service, 2007.

WORKING FIRE

1. Thanks to Pete Duncan and Dan Elliott for insightful conversations about the Plumas and its past. It should go without saying that their interpretation of the forest and its history differs from mine.

2. On hydraulicking I follow Grove Karl Gilbert, *Hydraulic-Mining Debris in the Sierra Nevada*, U.S. Geological Survey, Professional Paper 105 (Washington, DC: GPO, 1917).

3. On indigenous burning, see Kat Anderson, *Tending the Wild: Native American Knowledge and the Management of California's Resources*

(Berkeley: University of California Press, 2006). Quote from page 51, L. A. Barrett, "A Record of Forest and Field Fires in California from the Days of the Early Explorers to the Creation of the Forest Reserves" (unpublished manuscript, Pacific Southwest Research Station, 1935). On the general chronicle of fire, as registered by fire scars, see Tadashi J. Moody, JoAnn Fites-Kaufman, and Scott L. Stephens, "Fire History and Climate Influences from Forests in the Northern Sierra Nevada, USA," *Fire Ecology* 2, no. 1 (Spring 2006): 115–41, which expends most of its energy aligning fire data with climate data.

4. John B. Leiberg, *Forest Conditions in the Northern Sierra Nevada, California*, U.S. Geological Survey, Professional Paper 8, Series H, Forestry 5 (Washington, DC: GPO, 1902), 40–44, separate descriptions by drainage.

5. Leiberg, *Forest Conditions*, 41.

6. Leiberg quotes, *Forest Conditions*, 83–86 on Middle Fork, but fire is an informing theme throughout the survey. The Plumas survey quote is from Barrett, "Record of Forest and Field Fires," 48–49.

7. Barrett, "Record of Forest and Field Fires," 48. "Peculiar attitude" from *Fourth Biennial Report of the State Forester of the State of California* (Sacramento, CA: 1913), 11.

8. The most concise summary of the controversy can be found in Stephen J. Pyne, *Fire in America* (Seattle: University of Washington Press, 1997), 100–122.

9. Pat Terhune and George Terhune, "The Quincy Library Group," in *Engaging, Empowering, and Negotiating Community: Strategies for Conservation and Development* (workshop, Conservation and Development Forum, West Virginia University and the Center for Economic Options, October 8–10, 1998), 3.

10. Gary Snyder, *Back On the Fire: Essays* (Berkeley, CA: Counterpoint Press, 2008), 7–8.

THE PASTURES OF PURGATORY

1. Fr. Junípero Serra, *The Founding Document of Mission San Juan Capistrano* (n.p.: Fulton and Kay Shaw, 1976). For a good survey of California ranching, see L. T. Burcham, "California Rangelands in Historical Perspective," *Rangelands* 3, no. 3 (June 1981), 95–104.

2. Quotes from Barrett, "Record of Forest and Field Fires," 47, 22, except for Muir, which comes from John Muir, *The Mountains of California* (Garden City, NJ: Doubleday, 1961), 154.

3. I would like to thank Marc Nelson and his staff, in absentia, for the tolerance to absolve my error in giving the wrong date for a planned meeting and willingness, under very severe time constraints, to pick up some pieces. Special thanks to Rick Mowery for collecting and forwarding useful documents relevant to the Grindstone story. They are in no way responsible for my solo flight through the forest and pinched interpretations of its complex historical landscape.

4. From Samuel Trask Dana and Myron Krueger, *California Lands: Ownership, Use, and Management* (Washington, DC: American Forestry Association, 1958), 183–84.

5. See H. L. Shantz, *The Use of Fire as a Tool in the Management of the Brush Ranges of California* (California State Board of Forestry, 1947), 127, 123. Useful summary in Robert W. Cermak, *Fire in the Forest: A History of Forest Fire Control on the National Forests in California, 1898–1956*, R5-FR-003, U.S. Forest Service, California, July 2005, 295–99.

6. Cermak, *Fire in the Forest*. A. W. Sampson and L. T. Burcham, *Costs and Returns of Range Improvement in Northern California* (Sacramento: California Division of Forestry, 1954). Dana and Krueger, *California Lands*, 186.

7. Cermak, *Fire in the Forest*, 297; Dana and Krueger, *California Lands*, 186.

8. Mendocino National Forest, *Brushland Management the Grindstone Way* (unpublished report, 1980; revised, 1981).

9. H. H. Biswell, "Prescribed Burning in Georgia and California Compared," *Journal of Range Management* 11, no. 6 (1958): 293–98.

10. Chapter 11, in John Steinbeck, *The Pastures of Heaven* (New York: Penguin, 1986).

THE BIG ONES

1. I am the beneficiary of an extraordinary effort, organized by John Swanson and the East Bay Regional Parks, to educate me in the saga of fire in the Hills. I was privileged to hear analyses by Jerry Kent, Cheryl Miller, Ken Blonski, Rosemary Cameron, Leroy Griffin, and Peter Scott, and to visit the site in the field with John and Bill Nichols. A crowded day but an exceptionally instructive one, which left me with the belief that the only contribution I might make was to shift the terms of context. My thanks to them all. I was impressed with their resolve as much as with their knowledge.

2. I rely on Philip L. Fradkin's comprehensive *The Great Earthquake and Firestorms of 1906* (Berkeley: University of California Press, 2005); for figures see 9–10, 37.

3. Grove Karl Gilbert, April 20, 1906, Field Notebook 3501, National Archives. Quoted in Stephen J. Pyne, *Grove Karl Gilbert* (Austin: University of Texas Press, 1980), 211.

4. Jack London, "The Story of an Eye-Witness," *Collier's* 37, no. 6 (May 5, 1906): 22–25.

5. Figures from David J. Nowak, "Historical Vegetation Change in Oakland and Its Implications for Urban Forest Management," *Journal of Arboriculture* 19, no. 5 (September 1993): 313–19.

6. Many sources. I found most useful the National Fire Protection Association, *The Oakland/Berkeley Hills Fire*, report for the National Wildland/Urban Interface Fire Protection Initiative, 1992; and East Bay Hills Fire Operations Review Group, Office of Emergency Services, *The East Bay Hills Fire: A Multi-Agency Review of the October 1991 Fire in the Oakland/Berkeley Hills*, report for California Office of Emergency Services, February 27, 1992.

    The fire is the subject of numerous publications. I should mention in particular Margaret Sullivan, *Firestorm! The Story of the 1991 East Bay Fire in Berkeley* (City of Berkeley, 1993), and Peter Charles Hoffer, *Seven Fires: The Urban Infernos that Reshaped America* (New York: PublicAffairs, 2006), which includes several chapters on the Tunnel fire.

7. East Bay Hills Fire Operations Review Group, *East Bay Hills Fire*, 8.

8. Ibid., 7.

VIGNETTES OF PRIMITIVE AMERICA

1. I want to thank David Bartlett, Tony Caprio, Jan van Wagtendonk, Kelly Martin, and Brenda Lissoway for advising me on how fire behaved historically in the parks and for providing access to published information and archives. It was a wonderful, if far too brief, tour.

2. Bruce Kilgore, "Origin and History of Wildland Fire Use in the U. S. National Park System," *George Wright Forum* 24, no. 3 (2007): 103–4; Hal Rothman, *Blazing Heritage: A History of Wildland Fire in the National Parks* (New York: Oxford University Press, 2007), 112–13, 115–18. I am indebted to Tom Nichols for his perceptive comments on the institutional rifts that have plagued the NPS fire program from its origins.

3. For a good (and pictorial) distillation of those changes, see George E. Gruell, *Fire in Sierra Nevada Forests: A Photographic Interpretation of Ecological Change Since 1849* (Missoula, MT: Mountain Press, 2003).

4. Quotes from Bruce M. Kilgore, "Impact of Prescribed Burning on a Sequoia-Mixed Conifer Forest," 345–76, Peter H. Schuft, "A Prescribed Burning Program for Sequoia and Kings Canyon National Parks," 377–89, and John S. McLaughlin, "Restoring Fire to the Environment in Sequoia and Kings Canyon National Parks," 391–96, in *Proceedings, Tall Timbers Fire Ecology Conference, 1972* (Tallahassee, FL: Tall Timbers Research Station, 1973), specifically 372, 386–87, 389, 394–95.

5. Van Wagtendonk quote from conversation with author, July 2011.

6. A cogent version is available in Thomas M. Bonnicksen and Edward C. Stone, "Managing Vegetation Within U. S. National Parks: A Policy Analysis," *Environmental Management* 6, no. 2 (1982): 109–22. For the origins of the Christensen Report, see Norman L. Christensen et al., *Final Report: Review of Fire Management Program for Sequoia-Mixed Conifer Forests of Yosemite, Sequoia and Kings Canyon National Parks* (unpublished report, 1986), 2.

7. See Jan van Wagtendonk, "The History and Evolution of Wildland Fire Use," *Fire Ecology* 3, no. 2 (2000): 3–17.

8. A good summary history is available in Schuft, "Prescribed Burning Program."

EPILOGUE

1. Carl Wilson, "The People Pressures in Forest Fires Control," *Proceedings, Society of American Foresters, Denver, Colorado* (Washington, DC: Society of American Foresters, 1964), 59.

2. Mike Davis, *Ecology of Fire: Los Angeles and the Imagination of Disaster* (New York: Metropolitan Books, 1998), 6.

3. Kevin O'Leary, "Can Budget-Strapped California Afford More Wildfires?," *Time*, September 7, 2009, http://content.time.com/time/nation/article/0,8599,1920815,00.html.

4. Haas comment cited in David Wyatt, *Five Fires: Race, Catastrophe, and the Shaping of California* (Reading, MA: Addison-Wesley, 1997), 1.

INDEX

ABOUT THE AUTHOR

Stephen J. Pyne is a professor in the School of Life Sciences, Arizona State University. He is the author of over 25 books, mostly on wildland fire and its history but also dealing with the history of places and exploration, including *The Ice*, *How the Canyon Became Grand*, and *Voyager*. His current effort is directed at a multivolume survey of the American fire scene—*Between Two Fires: A Fire History of Contemporary America*, and *To the Last Smoke*, a suite of regional reconnaissances, all published by the University of Arizona Press.